Instrumentation

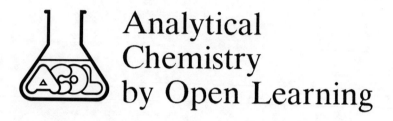

Analytical Chemistry by Open Learning

Project Director
BRIAN R CURRELL
Thames Polytechnic

Project Manager
JOHN W JAMES
Consultant

Project Advisors
ANTHONY D ASHMORE
Royal Society of Chemistry

DAVE W PARK
Consultant

Administrative Editor
NORMA CHADWICK
Thames Polytechnic

Editorial Board
NORMAN B CHAPMAN
*Emeritus Professor,
University of Hull*

BRIAN R CURRELL
Thames Polytechnic

ARTHUR M JAMES
*Emeritus Professor,
University of London*

DAVID KEALEY
Kingston Polytechnic

DAVID J MOWTHORPE
Sheffield City Polytechnic

ANTHONY C NORRIS
Portsmouth Polytechnic

F ELIZABETH PRICHARD
*Royal Holloway and Bedford
New College*

Titles in Series:

Instrumentation

Analytical Chemistry by Open Learning

Author:
GRAHAM CURRELL
Bristol Polytechnic, UK

Editor:
NORMAN B. CHAPMAN

on behalf of ACOL

Published on behalf of ACOL, London
by
JOHN WILEY & SONS
Chichester · New York · Brisbane · Toronto · Singapore

Published by permission of the Controller of
Her Majesty's Stationery Office

Library of Congress Cataloging in Publication Data:

Currell, Graham
    Analytical chemistry by open learning.
Instrumentation

    (Analytical chemistry by open learning)
    Bibliography: p.
    Includes index.
    1. Instrumental analysis—Programmed instruction.
2. Chemistry, Analytic—Programmed instruction.
I. Chapman, N. B. (Norman Bellamy), 1916–.
II. ACOL (Firm : London, England)   III. Title.
IV. Series.
QD79.I5C87   1987   543'.07   86-28141
ISBN 0 471 91368 5
ISBN 0 471 91369 3 (pbk.)   *66362*

British Library Cataloguing in Publication Data:

Currell, Graham
    Instrumentation.—(Analytical Chemistry)
    1. Instrumental analysis
    I. Title   II. Chapman, Norman B.
    III. Analytical Chemistry by Open Learning
    (Project)   IV. Series
    543'.08   QD79.I5

    ISBN 0 471 91368 5
    ISBN 0 471 91369 3 Pbk

Printed and bound in Great Britain

Analytical Chemistry

This series of texts is a result of an initiative by the Committee of Heads of Polytechnic Chemistry Departments in the United Kingdom. A project team based at Thames Polytechnic using funds available from the Manpower Services Commission 'Open Tech' Project have organised and managed the development of the material suitable for use by 'Distance Learners'. The contents of the various units have been identified, planned and written almost exclusively by groups of polytechnic staff, who are both expert in the subject area and are currently teaching in analytical chemistry.

The texts are for those interested in the basics of analytical chemistry and instrumental techniques who wish to study in a more flexible way than traditional institute attendance or to augment such attendance. A series of these units may be used by those undertaking courses leading to BTEC (levels IV and V), Royal Society of Chemistry (Certificates of Applied Chemistry) or other qualifications. The level is thus that of Senior Technician.

It is emphasised however that whilst the theoretical aspects of analytical chemistry can be studied in this way there is no substitute for the laboratory to learn the associated practical skills. In the U.K. there are nominated Polytechnics, Colleges and other Institutions who offer tutorial and practical support to achieve the practical objectives identified within each text. It is expected that many institutions worldwide will also provide such support.

The project will continue at Thames Polytechnic to support these 'Open Learning Texts', to continually refresh and update the material and to extend its coverage.

Further information about nominated support centres, the material or open learning techniques may be obtained from the project office at Thames Polytechnic, ACOL, Wellington St., Woolwich, London, SE18 6PF.

How to Use an Open Learning Text

Open learning texts are designed as a convenient and flexible way of studying for people who, for a variety of reasons cannot use conventional education courses. You will learn from this text the principles of one subject in Analytical Chemistry, but only by putting this knowledge into practice, under professional supervision, will you gain a full understanding of the analytical techniques described.

To achieve the full benefit from an open learning text you need to plan your place and time of study.

- Find the most suitable place to study where you can work without disturbance.

- If you have a tutor supervising your study discuss with him, or her, the date by which you should have completed this text.

- Some people study perfectly well in irregular bursts, however most students find that setting aside a certain number of hours each day is the most satisfactory method. It is for you to decide which pattern of study suits you best.

- If you decide to study for several hours at once, take short breaks of five or ten minutes every half hour or so. You will find that this method maintains a higher overall level of concentration.

Before you begin a detailed reading of the text, familiarise yourself with the general layout of the material. Have a look at the course contents list at the front of the book and flip through the pages to get a general impression of the way the subject is dealt with. You will find that there is space on the pages to make comments alongside the text as you study—your own notes for highlighting points that you feel are particularly important. Indicate in the margin the points you would like to discuss further with a tutor or fellow student. When you come to revise, these personal study notes will be very useful.

∏   When you find a paragraph in the text marked with a symbol such as is shown here, this is where you get involved. At this point you are directed to do things: draw graphs, answer questions, perform calculations, etc. Do make an attempt at these activities. If necessary cover the succeeding response with a piece of paper until you are ready to read on. This is an opportunity for you to learn by participating in the subject and although the text continues by discussing your response, there is no better way to learn than by working things out for yourself.

We have introduced self assessment questions (SAQ) at appropriate places in the text. These SAQs provide for you a way of finding out if you understand what you have just been studying. There is space on the page for your answer and for any comments you want to add after reading the author's response. You will find the author's response to each SAQ at the end of the text. Compare what you have written with the response provided and read the discussion and advice.

At intervals in the text you will find a Summary and List of Objectives. The Summary will emphasise the important points covered by the material you have just read and the Objectives will give you a checklist of tasks you should then be able to achieve.

You can revise the Unit, perhaps for a formal examination, by re-reading the Summary and the Objectives, and by working through some of the SAQs. This should quickly alert you to areas of the text that need further study.

At the end of the book you will find for reference lists of commonly used scientific symbols and values, units of measurement and also a periodic table.

Contents

Study Guide

Modern chemical instrumentation now uses very sophisticated measurement techniques. If you are to understand how, and why, these instruments work, it is essential to learn a few general principles of instrumentation science and technology. This Unit introduces fundamental concepts which have a wide application in analytical instruments. The Unit is not, however, a comprehensive catalogue of the different instruments to be found in a chemical laboratory. The physical description of particular types of instruments (eg spectrophotometers, thermal analysis instruments) can be found in Units devoted to each technique, and in the recommended text books.

It is the purpose of this Unit to introduce you to those vital aspects of general instrumentation science which significantly affect:

— the way measurements are made,

— the design of scientific instruments, and

— the sensitivity and accuracy of the instruments.

It will be assumed that you have some experience of using simple chemical instruments (eg a pH-meter) in the laboratory, and a knowledge of Physics up to at least GCE 'O' level. An understanding of chemistry equivalent to HNC in chemistry of the BTEC is not essential, although it would be helpful in providing the scientific context in which the instruments are to be used. A knowledge of electronics is not necessary.

The Unit is in five parts.

Part 1 provides the introduction and Part 2 the electrical background to the Unit. The introduction begins the discussion of the different categories of instruments that are used in the laboratory, and then Part 2 gives a brief review of some of the basic principles of electricity. The possible roles played by electricity in an instrument are also discussed, together with the basic techniques of electrical measurement. It is important, even if some of these basic topics may be very familiar to you, that you check that you can answer all of the questions correctly. It is the applications which are important and not just the fundamental theory.

Parts 3 and 4 introduce, in a sequential development, various important topics. The important concept of the Fourier Transform is introduced in a non-mathematical way in Part 3. It is then used to discuss the treatment of signals and information inside the instruments. Complicated electronic circuitry is avoided, although the student is introduced to the purpose and characteristics of some of the main circuit functions. Part 4 deals with the performance of some of the functional elements in instruments which affect the interpretation, presentation or passage of data. Part 5 brings together the topics introduced earlier and relates these to the final design and specifications of complete instruments.

The full importance of the subject matter of the text may not become totally apparent until you are faced with the problem of trying to understand the workings of a particular instrument as described in an operator's manual. For this reason, it is recommended that, while you are working through the text, you also read through any available instrument manuals or brochures, so that you can understand the relevance of this text to actual instruments. You are also advised, when you have finished working in detail through the Unit, briefly to re-read Part 1. This will then give you a framework for the subject area, from which you can progress to understanding the descriptions of instruments given in other Units or in other texts.

Bibliography

There is no book which covers exactly the material of this Unit. Most books concentrate on a physical description of the instruments for a range of different instrumental techniques. However, it is possible to find references in the following books to the concepts described in this Unit.

1. D.A. Skoog and D.M. West, *Principles of Instrumental Analysis*, 3rd edn, Holt Saunders, 1985.

2. H.H. Willard, L.L. Merritt, J.A. Dean, and F.A. Settle, *Instrumental Methods of Analysis*, 6th edn, Van Nostrand, 1981.

3. H.A. Strobel, *Chemical Instrumentation*, 2nd edn, Addison-Wesley, 1973.

Book 1 is certainly the most directly useful text and is presented in a readable manner. Book 2 has a very comprehensive coverage but can be difficult to read as an introductory text. Book 3 has a good introduction on electronics and is a readable text.

You will find that other books on aspects of 'instrumental analysis' will give descriptions of particular instruments. It is worth referring to these occasionally as the need arises.

1. Chemical Instrumentation

Overview

This introduction to the basic concepts of measuring instruments uses a functional-element approach, with the idea of a transducer expanded at this stage to signify the means by which a signal enters the instrument. The concept of an analytical instrument as an instrument in which a controlled experiment is performed is also presented. The principal types of analytical instruments are introduced in a general classification of the experimental processes used in these instruments.

1.1. INTRODUCTION

1.1.1. The Chemist and Chemical Instrumentation

There are many diffferent types of instrument used in the chemistry laboratory, to perform many different functions in a variety of ways.

∏   Think back to your previous experience in the laboratory and try to list three or four typical instruments:

1.

2.

3.

4.

Now ask yourself when and why you should use these instruments rather than a classical, non-instrumental, method.

There may be many different reasons for choosing an instrumental method instead of a classical method. For example, it may be quicker, cheaper, more accurate, more selective; it may make it possible to automate the measurements, or it may be the only means of solving a particular problem. However, very few instrumental methods are without their own disadvantages. The instrumental method may actually be slower, more expensive, less accurate, or less selective. Some criteria (speed, expense, etc) will appear on different sides of the argument in different circumstances.

Clearly there is no single clear-cut criterion by which one decides to use one instrumental method instead of either a classical method or an alternative instrumental method. The choice must be made by the analyst in the context of each particular problem. To be able to make that choice he/she must know enough about the different types of instruments to be able to make a sound assessment of their strengths and weaknesses.

How much then does the chemist *need* to know about the instruments?

It is unreasonable to expect someone who only wants to *use* an instrument to understand all about its inner workings. On the other hand, the chemist should not just be content to 'drive' the machine, without knowing the first thing about what the controls actually do. It is, however, essential to know under what conditions the results from a particular instrument may be reliable, or when they may actually be wrong, ie the *limitations* of the instrument. It is also important to be aware of the high degree of flexibility in many analytical instruments, and the different ways in which they are capable of being used under the control of an experienced operator, ie the *capabilities* of the instrument.

First of all, what do we mean by an *instrument*?

1.1.2. Concept of an Instrument

Many modern analytical instruments are extremely complex pieces of equipment. They may contain a great variety of component parts, including mechanical controls, optical and electronic systems, 'plumbing' for the flow of gases or liquids, radioactive sources and detectors, and recently, powerful computational electronics. This complexity is the result of combining many different aspects of instrumentation over a development period of many years.

The best way of beginning to understand modern analytical equipment is to look at the various stages of its development starting from the very simplest instrument, and discovering at each stage the different factors which contribute to current design. As a starting point, we can use a dictionary definition of an instrument as being:

'anything used as an aid in doing physical work'

There are two important words in this definition as far as chemical analysis is concerned. The first is that the instrument is only an *aid*. Despite the expense, the complexity, and the demands for expert maintenance made by many instruments, the instrument remains (or should remain) the *servant* of the analyst, to be used only when appropriate. Readings should never be taken blindly and without question. The analyst should be aware of all the conditions under which the instrument gives a measurement, and the consequent reliability of the result.

The other important word is *physical*. This may seem inappropriate for the instruments that are used in the chemical laboratory, but in fact all the analytical instruments in chemistry rely on some 'physical' process, eg the absorption of light, the measurement of temperature, the electrical conductivity of a solution. These physical properties then give the analyst clues (ranging from direct evidence to mere suggestion) which help to solve the chemical problem. Our next step is to appreciate how an instrument can actually be an aid to the analysis.

1.1.3. Experiment, Measurement, Interpretation, and Analysis

A crucial aspect of most analytical procedures is the act of *measurement* itself. It is at this point that instruments first come into use. They extend the five senses of the analyst so that it is no longer necessary to rely entirely on sight, hearing, smell, touch, or taste. Many other quantities are now capable of being measured, and their values recorded. For example in chemistry we might measure pH value, electrical conductivity, refractive index, temperature.

The concept of Measurement Instrumentation is not limited to chemical analysis but covers all aspects of our technological world. We have, for example:

> thermometers to measure temperature,
> barometers to measure pressure,
> hygrometers to measure humidity,
> accelerometers to measure acceleration,
> anemometers to measure wind speed,
> light-meters to measure light intensity.

The list is almost endless!

The output from these instruments is directly related to the value of the quantity being measured, and there are very few (if any) operator controls. These are instruments for measurement only. We would not classify them as analytical instruments. However, when performing an analysis, a chemist will often need to employ measurement in association with some form of *experimental procedure* in order to separate and/or identify the components of a system. It will also be necessary to *interpret* the results of the measurement that he has made.

Hence:

$$
\left.\begin{array}{c}
\text{Experiment} \\
+ \\
\text{Measurement} \\
+ \\
\text{Interpretation}
\end{array}\right\} \text{ leads to Analysis}
$$

Clearly, measurement instrumentation must form an important part of any instrumental analysis, but we shall see later how the *Analytical Instrument* of the chemist has been specifically developed so that it does more than simple measurement. The analytical instrument also performs a significant part of the experimental procedure itself, and with modern computational systems frequently carries out much of the interpretation.

We start our study with at look at the elements which go to make up a simple Measuring Instrument.

∏   In the third century BC, Archimedes was required by Hieron II, King of Syracuse, to determine whether a new crown was pure gold or a mixture of gold and silver. Archimedes realised that if he could measure the density (from mass and volume) of the material, then, knowing the densities of gold and silver, he could calculate how much silver had been added to the gold. To measure the volume of the crown he hit on the idea of finding how much water was displaced when the crown was completely immersed.

Show how the analysis performed by Archimedes contained the three components: experiment, measurement, and interpretation.

Experiment.

Here, the experiment involved the use of the water displacement technique to equate the volume of the crown (of complicated shape) to a volume of water which could be easily be measured.

Measurement.

The quantities measured by Archimedes were the mass of the crown (gravimetric measurement) and the volume of the water displaced (volumetric measurement).

Interpretation.

The density of the crown was calculated from these two measurements. The densities of pure gold and silver were already known. The decision on the content of the crown was achieved by comparing the measured results with known facts.

1.2. ELEMENTS OF A MEASURING INSTRUMENT

1.2.1. Introduction

In order to illustrate the elements of a simple measuring instrument, we shall use two examples – an ordinary thermometer and a lightmeter. These examples of instruments, together with others that will appear in later sections, will be used to illustrate the important aspects of instrumentation. You will probably have to refer back to these descriptions when reading later text or attempting to answer questions.

1.2.2. Mercury-in-glass Thermometer

The thermometer is used to measure temperature. When the temperature increases, the glass bulb at the end of the thermometer remains almost constant in volume, but the mercury expands – Fig. 1.2a.

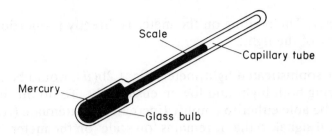

Fig. 1.2a.   *Mercury-in-glass thermometer*

This increase in volume of the mercury is very small and would normally be almost undetectable, except that the expansion has to take place in the very narrow bore of the capillary tube in the stem of the thermometer. A small expansion is now visible as an increase in the length of the mercury column along the thermometer scale.

1.2.3.   Light-meter

A light-meter measures light intensity, and may be used, for example, to measure the brightness of daylight for photography. The simple light-meter consists of a photo-detector and a meter – Fig. 1.2b(i).

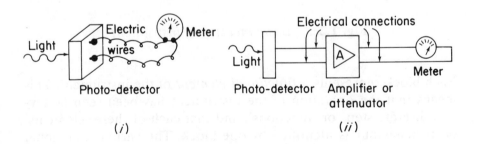

Fig. 1.2b.   *Light-meter*

The photo-detector is a device which will produce an electric current when it is illuminated. The magnitude of the current is proportional to the intensity of the light; this current is used to operate the meter. The reading on the meter is directly proportional to the intensity of the light.

A more sophisticated light-meter, Fig. 1.2b(*ii*), would be capable of measuring both high- and low-intensity light. To do this it is necessary to be able either to amplify (increase) or attenuate (reduce) the electrical signal so that it remains 'on scale' on the meter. This extra electronic circuitry would then alter the 'sensitivity' of the meter.

1.2.4. Block Diagrams

It is common practice to use *block diagrams* to represent the operation of an instrument. The first and simplest of these is given in Fig. 1.2c to represent a simple basic measuring instrument.

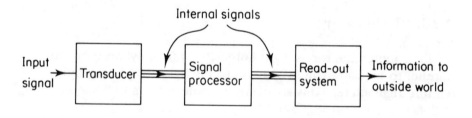

Fig. 1.2c. *Basic measuring instrument*

Each block represents a *functional element* of the instrument. This means that the operation of the instrument has been seen as several discrete steps or 'functions', and that each of these 'elements' of its behaviour is identified by one block. The various functional elements are more fully described in the following parts of this section. The purpose of this description is to clarify *understanding* of the instrument by breaking it down into manageable units. It is not intended to give a photograph of the inside of the instrument. Indeed, a photograph might show only a mass of wires, electrical components, motors, etc, with little obvious relation to each other, and be of little use to someone trying to understand basic concepts.

At this stage we are interested in 'what it does' rather than 'how it does it'.

1.2.5. Signal

An important and frequently used term in all instrumentation is *signal*. This word is used in very many different contexts. However, without going into the realm of Information Theory, we can say that, usually, a signal either represents some form of *information*, or it is a means of *carrying information*. For example, we humans communicate (exchange information) by some form of signal – by speech, writing, gestures, facial expressions.

The light-meter will use the intensity of light falling on it as its source of information. The photo-detector responds to the light signal and generates a proportional electric current – a new signal. This new signal carries the information giving the brightness of the light. In this example, the 'information' is originally in the form of a light signal, but this information is then transferred to an electrical signal.

1.2.6. Input-signal

This is the quantity being measured. For the light-meter, the 'input-signal' is the intensity of light falling on the detector. For the thermometer, the 'input-signal' is the temperature of the substance. Thus the input signal may represent:

(*a*) actual energy entering the system (eg from light intensity),

(*b*) the value of some parameter being measured (eg temperature).

In an Analytical Instrument the input-signal is sometimes called the *Analytical Signal*.

∏   What is the input-signal for each of the following?

A barometer,

a hygrometer,

an accelerometer,

an anemometer.

The answers are listed below.

INSTRUMENT	INPUT-SIGNAL
barometer	pressure
hygrometer	humidity
accelerometer	acceleration
anemometer	wind speed

These answers could have been found by looking at the list of measuring instruments given in Section 1.1.3.

The 'input-signal' is the quantity which the instrument is measuring. In all the examples given in the question, the input signal is the *value* (information) of some parameter of the system and is *not* in the form of *energy* flowing into the instrument. For a light-meter, however, the input-signal *is* in the form of an energy input. Hence, an input-signal may be either information in the form of the magnitude of some quantity (eg pressure), or it may be a real flow of energy (eg light) into the detector.

1.2.7.   Transducer

'Transducer' is the general name given to a functional element which receives the input-signal and gives an output-signal which is directly related to the magnitude of the input-signal. With reference to our simple example of the thermometer – if we take the mercury to be acting as the transducer, then the *input-signal* is the *temperature* of the substance, and the *output-signal* is the *expansion* of the mercury. Other transducers will give an output-signal in the form of an electrical signal. The photo-detector in the light-meter is an obvious example in which the light intensity is the input-signal and the electric current is the output-signal.

All transducers work through some form of physical process. In the thermometer, the fluid in the tube *expands* if the temperature (the input-signal) increases. In the photo-detector the process is more difficult to understand, but in general terms the light gives energy to the electrons in the detector and these electrons will then flow through an external circuit to give an electric current. Some transducers are extremely complex systems themselves. In fact the development of special transducers is the next stage of development in chemical instrumentation. We discuss this in Section 1.3.

1.2.8. Read-out System

The transducer (above) is the functional element at the input to the instrument.

At some point, the measuring instrument must communicate its results to the outside world – an output. In the *simplest* situation this will be a visual indication, such as the change in length of the mercury column in a thermometer, or the deflection of the needle in a meter recording the output from photo-detector. Alternatively, the result may be fed into a chart recorder to give a permanent record on paper, or it may be stored on magnetic tape.

There are many other ways in which the final result may be produced (see Fig. 1.2d), but they can all be classified under the heading 'read-out system'. An increasingly common method feeds the results, still in an electronic format, directly into a computing facility or data station.

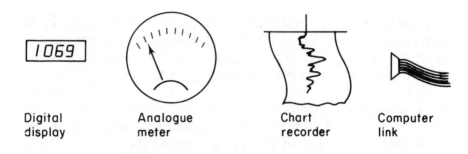

| Digital display | Analogue meter | Chart recorder | Computer link |

Fig. 1.2d. *Some read-out systems*

1.2.9. Signal Processor

The output-signal from the transducer is often not in the correct form to be connected directly to the read-out system. For example, in a particular instrument the transducer may give only a few millivolts output whereas the read-out device may need 5 volts to operate. An electronic amplifier between the transducer and read-out system will be required to increase the voltage of the signal. Alternatively, the electrical signal from a simple light detector may be in the form of an electric *current* and this must be converted into a *voltage* signal before reaching the read-out. The functional element needed to do this is called a 'current-to-voltage' converter.

The 'go-between' that transforms the output-signal from the transducer so that it is acceptable to the read-out system, is called a '*signal processor*'. The signal processor may take many different forms in electronics, of which two examples have been mentioned above.

It may also be of a non-electronic nature. For a simple thermometer, we have identified the transducer as the mercury which undergoes an expansion. In fact this expansion is very small indeed and would normally be quite invisible to the naked eye. However, the shape of the glass of the thermometer forces the expansion of the mercury to take place along a very narrow tube thus 'amplifying' the physical movement of the mercury in expansion. The 'shape' of the thermometer could be interpreted as acting as a 'signal processor'. The 'read-out system' is then the position of the mercury in the narrow tube against the calibrated scale.

∏   Indicate, for each of the following statements, whether it is
     true or false.

   (a) A signal processor is used to convert light intensity into
        an electrical signal which operates the read-out system.

                                                        T / F

   (b) A transducer converts an analytical signal into a signal
        of a different form.

                                                        T / F

   (c) A read-out system amplifies a signal.

                                                        T / F

   (d) It is possible to have a measuring instrument which
        does not have a signal processor as one of its functional
        elements.

                                                        T / F

The answers are as follows.

(a) F   A signal processor modifies a signal (eg amplification),
         but normally the *form* of the signal is not changed. If the
         signal entering the processor is electrical, then the sig-
         nal leaving will also be electrical. Similarly a mechanical
         signal may be modified by a signal processor, but it will
         remain in a mechanical form.

(b) T   The transducer is the element which converts the ana-
         lytical signal from one form (eg electrical, optical, me-
         chanical) into a different form.

(c) F   According to our simple definition the read-out system
         does what it says: it provides a way of reading the result
         of the measurement. The signal-processor is the func-
         tional element which amplifies the signal.

(*d*) T   The signal processor is not needed if the output from
the transducer is already capable of operating the read-
out system directly – eg the photo-detector which can be
connected directly to a meter.

SAQ 1.2a

A simple aneroid barometer is a measuring
instrument designed to measure pressure (see
Fig. 1.2e). It consists of flexible sealed bellows,
which expand if the atmospheric pressure drops,
and contract if the pressure rises. A mechanical lever-and-gear system operates a movable
pointer which gives a reading of pressure.

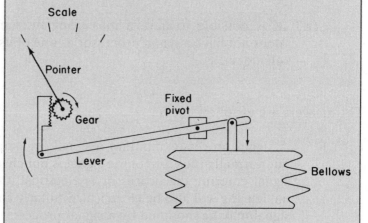

Fig. 1.2e.   *Aneroid barometer*

Apply to this instrument the diagrammatic representation of a measuring instrument that was
introduced in the text, and explain which parts
of the barometer are represented by the following:

(*i*)   transducer,

(*ii*) signal processor,

(*iii*) read-out system.

SAQ 1.2a

1.3. TRANSDUCERS AND DETECTORS

1.3.1. Role of the Transducer or Detector

We introduced the idea of a transducer in the previous section. It is a device which will respond to an input-signal, and will give an output-signal whose magnitude is related to that of the input-signal.

It is used for converting either *energy* or *information* from one form into another.

— The photo-detector converts light intensity into electric current (proportional to the light intensity).

— The bellows in an aneroid barometer converts changes in the magnitude of atmospheric pressure into a mechanical movement.

There are many examples of transducers used in all areas of our daily lives.

A very significant advance in instrumentation has been the development of transducers specifically designed to help the chemist. We still use the idea of a Measuring Instrument as in Section 1.2, but, by making transducers capable of responding to specifically chemical variables, we have now our first category of 'chemical' instrumentation.

One of the most useful examples is in electrochemistry where we can consider the electrode as a transducer. The most common example is that of the pH electrode which gives an electric voltage (output-signal) whose magnitude depends on the concentration of hydrogen ions in solution (input-signal). There is also an increasing range of ion-selective electrodes, ie electrochemical transducers which have been developed to respond selectively to different input-signals, each representing the concentration of a particular ion in solution.

The development of transducer technology is not confined to electrochemistry, but is rapidly expanding in many areas. Another example is in gas detection, where small quantities of specific gases change the electrical conductivity (and therefore the flow of electricity) in thin organic films. These transducers respond selectively to particular gases.

You will find that the term 'detector' is used frequently in place of 'transducer'. Do not worry about which word to use – they are now almost interchangeable in common usage. A transducer usually involves a measurement of the magnitude of the input-signal, whereas a detector may give only an 'on/off' response depending on whether the input-signal is above or below a critical level – eg a smoke detector. Nevertheless, the term 'detector' is now also applied to most transducers found in chemical instrumentation.

An important feature of transducers (or detectors) for analysis is obviously their *selectivity*, ie the extent to which they will respond only to a given input-signal. A sodium-ion electrode has a high 'selectivity' if it responds only to the sodium ions in solution, and will not give a response for any ions other than sodium in the solution. We shall discuss *selectivity* together with some of the other specifications for transducers in Section 1.3.4, but first we examine two examples of transducers (detectors) used in different fields of analysis.

1.3.2. The Combination pH-Electrode

The physical and chemical processes occurring in the combination pH electrode are very complex, and it is not appropriate to attempt a detailed discussion here. It is sufficient for our purposes to know that the part in the electrode, that is sensitive to pH, is a thin membrane of special glass which separates the solution being tested from a saturated solution of silver chloride, usually in 0.1 M–HCl held inside the electrode. A voltage difference is generated between the two surfaces of the glass membrane, and the magnitude of this voltage depends on the pH of the solution tested – see Fig. 1.3a(i).

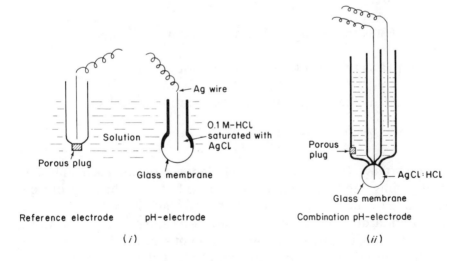

Fig. 1.3a. *pH Measurement*

In order to measure this voltage difference there must be electrical contacts on each side of the membrane.

Inside the electrode the contact is made to the silver chloride solution by using a silver wire. This is the 'pH' electrode. The electrical 'contact' to the solution being tested is made by using another electrode called a *reference* electrode. This has the special feature that it is not itself sensitive to changes in the pH of the solution (otherwise this would confuse the results from the glass membrane).

In combination pH-electrodes, the 'pH' and 'reference' electrodes are combined into one unit: see Fig. 1.3a(*ii*). The theoretical voltage output from a combination pH-electrode (the voltage difference between the 'pH' and the 'reference' electrode) is given by the equation, originally derived by Nernst:

$$E_v = E_o - 0.198 \times T \times (\text{pH} - 7.0)$$

where

E_v is the voltage being measured in millivolts,

pH is the pH of the solution under test,

T is the temperature of the solution in Kelvin

(25 °C = 298 K).

E_o is a voltage (in millivolts) which depends on the particular electrode being used. It is almost constant, but will change from one electrode to another and will also change gradually with time. When you 'standardise' a pH meter, you are correcting for different values of E_o by using a standard buffer solution of known pH and adjusting the 'set buffer' control until the meter reads the correct value.

The theoretical values for E_v can be calculated (except for E_o) for different situations. For example:

at 25 °C and pH = 9.00,

with $T = 273 + 25 = 298$ K

$E_v = E_o - 0.198 \times 298 \times (9.00 - 7.00) = (E_o - 118)$ mV

at 30 °C and pH = 4.00,

with $T = 273 + 30 = 303$ K

$E_v = E_o - 0.198 \times 303 \times (4.00 - 7.00) = (E_v + 180)$ mV.

We have seen that a pH-electrode is a source of voltage (*ca.* 0–0.4 V). The same can be said of a car battery (12 V) and the battery for a calculator (1.5 V). However, we can measure the voltage of the two batteries by using a very simple electrical meter.

∏   Why do we need a special meter to measure the voltage of a pH-electrode?

The difference is that a car battery can provide up to several hundred amperes of electrical current, and a calculator battery can provide up to 0.1 A, which, in both examples, is more than enough current (*ca.* 50 microamps) to 'drive' a simple meter. However, a pH electrode is unable to supply even one microamp (millionth of an amp) – it is a source of voltage which is virtually without current. The pH meter is specially designed to include an electronic system which is capable of measuring voltage without taking any current (or very little) from the electrode itself.

During this introduction to the pH-electrode we have been talking about 'voltage'. You will find that when you refer to the text book, the term used will be 'potential difference'. Do not worry at this stage about the difference (potential difference is measured in volts!). This will be mentioned later in Section 2.1.

∏   Calculate the change in the voltage produced by a pH-
         electrode at 25 °C when it is moved from a solution of pH
         4.0 to a solution of pH 9.0.

At 25 °C, $T = 273 + 25 = 298$ K.

At pH 4.00,

$$E_v = E_o - 0.198 \times 298 \times (4.00 - 7.00) = E_o + 177 \text{ mV}.$$

At pH 9.00,

$$E_v = E_o - 0.198 \times 298 \times (9.00 - 7.00) = E_o - 118 \text{ mV}.$$

The difference between these two values is,

$$E_o - 118 - (E_o + 177) = -295 \text{ mV}.$$

The voltage will *decrease* by 295 mV.

1.3.3.   Thermal Conductivity (Hot-wire) Detector

A Thermal Conductivity Detector, TCD, is used in gas–liquid chromatography to detect the presence of a sample of volatile organic
material flowing along with a 'carrier' gas. The detector consists of
a wire filament through which an electric current passes. The filament gets hot in the same way as the filament of a light bulb. The
temperature that the filament reaches depends on how quickly the
heat is being conducted away through the gas surrounding it: rapid
conduction of heat will result in a relatively low temperature, and
poor conduction of heat will give a relatively high temperature, see
Fig. 1.3b.

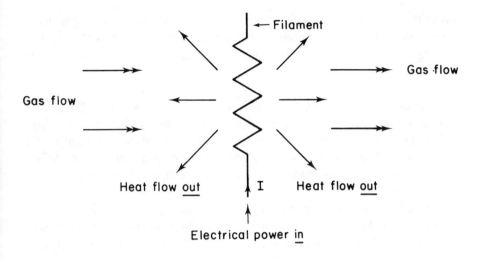

Fig 1.3b. *Gas flow and filament in a TCD*

The carrier gas (hydrogen or helium) has a high thermal conductivity and, when it is passed round the filament, the filament will reach an equilibrium temperature, T_0. When a volatile organic compound is mixed with the gas, the combined thermal conductivity of the carrier gas, plus sample, is less than that of the pure gas, the heat flow is reduced, and the temperature of the filament increases to a temperature, T_s. The presence of the sample (the input-signal) is converted into a change in temperature,

$$\Delta T = (T_s - T_0) \text{ of the filament.}$$

However, this is not the end of the story. How do we measure the temperature of the filament? The filament is made of fine platinum, gold, or tungsten wire, or alternatively, it is a semi-conducting thermistor. All of these have the property that if their temperature changes, then so does their electrical resistance. We now become interested in a change in resistance,

$$\Delta R = (R_s - R_0).$$

This is still not the end of the story, because we find that the rate of flow of the carrier gas also affects the rate at which heat is conducted away from the filament. Hence, a fluctuation in the carrier-gas flow-rate may still cause a change in the temperature, T_o (and thus in the resistance, R_o), even if there is no sample gas mixed with the carrier gas.

This problem can be solved by splitting the initial flow of gas into two streams, and passing each stream over a separate similar filament. Only one stream carries the sample. We now use an electric circuit (see Fig. 2.4g) which is capable of responding to the *difference* between the resistances, R_s and R_r, of the two filaments.

A variation in gas flow-rate will affect the resistance, R_o, of both filaments in the same way, and thus its effect on the final result will cancel out. This leaves only, ΔR, the difference due to a sample being carried in one of the gas streams.

$$\text{Difference} = R_s - R_r = (R_o + \Delta R) - R_o = \Delta R$$

The filaments are already carrying a current (the heater current, I) and so a change in the resistance, ΔR, is immediately converted into a change in voltage,

$$\Delta V = I \times \Delta R$$

The output from the detector is therefore given as a change in voltage.

∏   What would happen to the temperature of the filament of a TCD detector if the carrier gas stopped flowing?

If the flow of carrier gas stops, then the rate at which heat is conducted away from the filament *decreases*. Thus the *temperature* of the filament will rise. In some detectors, if the flow of carrier gas stops, the rise in temperature will be so great as to cause the filament to 'burn-out'. This is a practical point needing care when one uses a TCD!

Some TCD systems have an automatic safety device which will prevent permanent damage to the filaments if the flow stops.

1.3.4. Characteristics of a Transducer/Detector

The 'characteristics' of a component part of an instrument, are given as a list of concise statements about what that part can do. The characteristics are designed to give a clear and accurate indication of the performance of the unit.

For a transducer, some of the parameters in which we would be interested are given below.

Input-Signal.

> This is the quantity that the transducer is designed to measure.

Output-Signal.

> This gives the form of the output-signal. For example, this may be a voltage (as with the pH-electrode) a current (photo-detector), or mechanical movement (barometer bellows).

Quite clearly, the form of the output signal from the transducer will affect the type of signal processor that must be used to convert the signal into a form and magnitude that will operate the read-out system.

Sensitivity.

> Sensitivity is defined by the ratio below.

$$\frac{Magnitude\ of\ the\ output\text{-}signal}{Magnitude\ of\ the\ input\text{-}signal}$$

This ratio will normally have mixed units.

For example, for a light detector the input could be measured in lux (units of light-intensity) and the output measured in amperes (units of electric current), thus the sensitivity could be given in units of amperes lux^{-1}.

Selectivity.

> As explained in Section 1.3.1 this is concerned with the abil-
> ity of the transducer to respond only to the quantity required,
> and to ignore all others. Thus selectivity is given by the ratio
> below.

$$\frac{\textit{Sensitivity to the input-signal to be measured}}{\textit{Sensitivity to other input-signals}}$$

We are normally concerned with the ability of the transducer to
distinguish between two different quantities. For example. we may
want to know how successful use of the sodium-selective electrode is
in measuring only the concentration of sodium ions in the presence
of potassium ions. However, many detectors are not very selective.
The response of the TCD detector, for example, depends only on
the thermal conductivity of the sample; if two different samples have
the same thermal conductivity, then, if they arrive at the detector *at
the same time*, they cannot be distinguished by the detector.

There are, of course, many other important parameters such as the
signal-to-noise ratio and the stability. We shall discuss some of these
in later sections.

∏   Detectors, such as the TCD and the photo-detector, are not
     selective – they cannot themselves easily distinguish between
     different compounds. How then can they be used in analysis?

The answer to this question anticipates the work of later sections,
but you may have obtained a clue from Section 1.1.3 in which we
said that analysis will contain an experiment as well as a measurement. If the transducer is not sufficiently selective to isolate exactly
what we want to measure, then we can use an 'experiment' either
to isolate a particular variable (see Section 1.1), or physically to
separate different parts of the sample before measurement. If you
look back at 'selectivity' above, we hint that a TCD could be used
to measure the amounts of *different* constituents of a sample, only
if it can measure them *at different times* – this forms the basis of
chromatography.

1.3.5. Conclusion

In this section we have begun to study the use of special transducers or detectors in the field of chemical analysis. Some of these transducers are very complex systems themselves, although we can often treat them as self-contained units, provided that we know what goes in (input-signal) and what should come out (output-signal). We should always be aware of the physical processes taking place within the transducers, because this will help us to be aware of possible errors that may be introduced into our measurement by the instruments we are using.

It is useful to note here, that some text books extend the concept of the transducer almost to cover the complete instrument. A spectrophotometer is sometimes referred to, in effect, as an enormous transducer by which one measures the absorption of light by a chemical solution. The reason for this is a desire to make all instruments fit into the diagrammatic representation of a Basic Measuring Instrument given in Fig. 1.2c, regardless of how unbalanced this representation becomes. However, in this text the term transducer is restricted to the actual self-contained device which responds directly to one physical quantity – for example, in the spectrophotometer the transducer is just the light detector and not the whole instrument.

We shall develop in the next section the concept of an Analytical Instrument in its own right, with its own diagrammatic representation derived from that of the basic Measuring Instrument.

SAQ 1.3a   If the expression for the voltage from a pH electrode is rewritten as

$$E_V = E_0 - A\,(\text{pH} - 7.00)$$

calculate:

(i)   the value of A when the temperature of the solution is:

(a) 20 °C, (b) 25 °C, (c) 30 °C;

(ii)   the change in voltage if the pH of the solution changes by one pH unit (eg from 3.00 to 4.00) for each of the temperatures:

(a) 20 °C, (b) 25 °C, (c) 30 °C.

SAQ 1.3b

Assume that a filament in a TCD is made of a semi-conducting material, whose resistance decreases with increase of temperature. The filament is heated by the passage of a constant electric current, and pure helium gas flows over it. Some volatile organic material is added to the helium (without changing the flow rate). Indicate below whether the following parameters will increase or decrease.

Filament temperature: Increase / Decrease

Filament resistance: Increase / Decrease

Filament voltage: Increase / Decrease

SAQ 1.3c   List the Input-Signals and Output-Signals for
            each of the following:

            (*a*) photo-detector,

            (*b*) pH electrode,

            (*c*) TCD.

1.4. THE ANALYTICAL INSTRUMENT

1.4.1. Introduction

In Section 1.2 we looked at what we mean by a basic *measuring
instrument*. We can now consider a further development in instrumentation for the chemical analyst.

In performing an analysis, the chemist usually has to carry out an experiment. As an example, this 'experiment' could be a titration. The end-point of the titration might well be detected by using some form of instrument, measuring, for example, pH or conductivity.

The next stage in instrument development is when the instrument actually carries out part (or all) of the experiment on behalf of the analyst, the instrument then not only makes measurements where appropriate, but performs certain tasks which help in the experimental procedure. For example, in a titration, it is possible for an instrument to:

— to control (by using valves) the flow of liquid (titrant) from a burette,

— to detect the end-point by measurement as above,

— to 'tell' the operator the amount of titrant required.

This type of instrument therefore contains some form of *analytical procedure*, and could be called an *Analytical Instrument*. An actual experiment is performed within the instrument, and the results of this experiment are measured and presented by the instrument as part of a complete procedure.

Fig. 1.4a gives a block diagram of the *functional elements* of a general Analytical Instrument. You can see that the basic *measuring instrument* that was discussed in Section 1.2 now forms the 'Detection System' of a more complicated *analytical instrument*.

We shall discuss the individual 'functional elements' in Section 1.4.4, but first, in order to help your understanding of the functions that operate within an analytical instrument, we now introduce descriptions of two further instruments: the melting point apparatus, and the colorimeter. We then proceed to use these examples in later text.

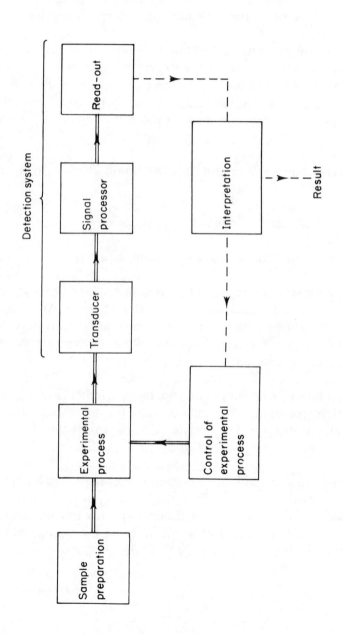

Fig. 1.4a. *Analytical instrument*

1.4.2. Melting-point Apparatus

The melting-point apparatus (Fig. 1.4b), as its name suggests, measures the temperature at which substances melt.

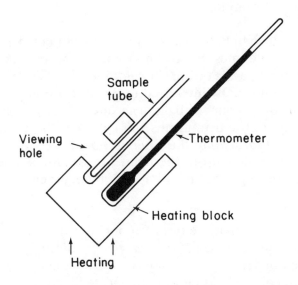

Fig. 1.4b. *Melting-point apparatus*

A small dry sample of the substance being examined is placed in a thin-walled glass tube which is inserted into a recess in a metal heating-block. The temperature of the block is raised by means of an electric heating coil (or a gas flame), and the point at which the sample melts is observed by eye. The corresponding temperature can be read from a thermometer which is in contact with the block. Note that the thermometer, which was described as a measuring instrument in Section 1.2 is now a component part of a more complicated analytical instrument.

The operator of the instrument is able:

— to insert and remove samples,

— to control the rate of heating of the block,

— to observe the temperature of the block.

1.4.3. Colorimeter

A colorimeter (Fig. 1.4c) is a simple instrument which measures the absorbance of light when it passes through a container (cuvette) containing a liquid sample.

A simple low-voltage bulb is used as the source of light. This is very similar to, and sometimes exactly the same as, an ordinary torch bulb. The light then passes through a *coloured filter*. The reason for this is that we are normally interested in the absorption of a particular wavelength or 'colour' of light. The filter absorbs the colours that are not required and transmits the colour to be used. (Note that the colour *absorbed* by the filter is *not* the colour you see when you hold the filter up to the light. For example, a red filter will let through red light but it will absorb blue, green, and yellow light.)

Later on we shall find that the term 'colour' is not sufficiently precise, and we shall use the idea of the *wavelength* of the light. This is, in fact, reflected in the name of the instrument. A colorimeter uses filters to obtain 'colours' but more sophisticated instruments, called spectrophotometers select particular wavelengths of light from the electromagnetic spectrum.

The light now passes through the liquid which may be held in a simple test tube or a specially made rectangular container called a cuvette. Depending on the concentration of the sample in solution, the intensity of the light will be reduced by absorption of some of the light energy by the sample.

Note that, as above, if the sample appears to be red in colour for example, then it is *not* absorbing very much red light. It appears red because the other colours in the 'white' light have been absorbed more than red. Therefore, for a 'red' solution we are not interested in the absorption of red light, but in the absorption of blue and green light. Consequently, it would be incorrect to use a red filter when the solution is also red in colour. We should have to use a filter of a complementary colour, ie a blue-green filter.

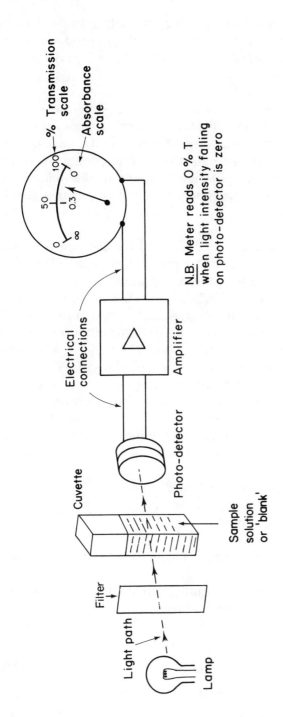

Fig. 1.4c. *Colorimeter*

After it has passed through the sample, the light falls on a photo-detector. The photo-detector is a device which will produce an electric current when it is illuminated. The magnitude of this current is proportional to the intensity of the light.

However, the magnitude of the current from the photo-detector may be very small, and may be insufficient to operate the indicating meter on the outside of the instrument. Consequently, an electronic amplifier must be used to increase the current produced, so that it is enough to drive the output-meter. Note that the 'light-meter' from Section 1.2, is now a component part of a complete analytical instrument.

The scale of the output meter is calibrated for each measurement by first inserting a cuvette containing pure solvent (the 'blank') and adjusting the amplification (or gain) of the amplifier, so that the meter reads 100% transmission of light for the pure solvent. The blank is then replaced by the sample which absorbs some of the light, the current to the meter drops, and the meter indicates a transmission of light on a scale between 0% and 100%. The meter reading gives the percentage transmission of light through the sample. You will find (Fig. 1.4c) that in addition to a percentage transmission scale (or sometimes instead of it), there may be an 'absorbance' scale which goes from 'Infinite absorbance' (0% transmission) to 'Zero absorbance' (100% transmission). The significance of this scale will be given in the unit on spectroscopy.

The operator of the instrument is able:

— to insert and remove samples;

— to change the filters and hence change the colour being used;

— to change the amount of amplification of the electrical signal from the detector to the output meter.

1.4.4. The Analytical Instrument

The principal functions (Fig. 1.4a) within an analytical instrument are as follows.

An Experimental Process.

There is a very wide range of experimental processes used in different instruments, eg absorption of light by a solution at different wavelengths; electrical conduction in a solution through an electrode at different potentials; weight loss of a substance during heating.

Control of the Experimental Process.

The essential feature of an analytical instrument is that the experimental process is, to some extent, under the control of the operator, eg choice of wavelength, rate of heating of sample.

Analytical Variables.

As a result of the experimental process, certain variables may be available for measurement, for example, the intensity and wavelength of light, temperature, time, distance. By interpreting the measured values of these variables, it is possible to achieve the necessary isolation or identification of the sample.

Detection Systems to Record Analytical Quantities.

These detection systems are essentially measurement instrumentation as discussed in the previous section. They form a component part of the analytical instrument. Obvious examples are the measurement of light intensity and temperature.

Sample Preparation.

Just as in any other experiment, the preparation of the sample is crucial. For example, it may be necessary to dilute a concentrated sample, or even to carry out a preparatory chemical process to obtain the sample in the correct chemical form, so that it is sensitive to the experimental process in the instrument. Details of sample preparation are given in a special unit on the subject, as well as in the individual subject units.

Interpretation.

The function of interpreting the results is part of the overall analytical process, but it may or may not be part of the instrument. The interpretation of the readings from an instrument may sometimes affect the way in which the operator chooses to control the instrument. For example, having made a rough measurement, the operator may choose to repeat a measurement which concentrates on a limited range of control settings to increase the precision of measurement. In modern instruments, sophisticated microelectronics now perform many of the interpretation functions that were previously done outside the instrument by the analyst.

However simple it may be to operate a modern sophisticated instrument, the operator must always realise that an experiment is being conducted on his behalf. It is the duty of the operator to be fully aware of the conditions of this experiment if he is to make responsible use of the results.

∏   Describe how each of the principal functions listed above for an analytical instrument can be related to a melting-point apparatus.

We can see how the functions of a general analytical instrument relate to the melting-point apparatus.

Experimental Procedure.

The temperature of a small sample is made to rise slowly, and the point at which the sample melts is observed by eye. The corresponding temperature is read from a thermometer.

Control.

The operator can, with practice, control the rate of change of temperature so that the exact melting point can be judged accurately. What often happens is that an inexperienced operator will start with a very small heating-rate at a temperature well below the actual melting-point, and, being impatient, will gradually increase the heating-rate so that, by the time the melting-point is reached, the temperature is rising so quickly it is difficult to measure the temperature accurately. As in many types of measurement it is often better to do a rough and quick preliminary measurement to get an approximate value, and then return to near the correct temperature and perform a slow and accurate measurement. This is an example of the way in which 'interpretation' can affect the way in which the instrument is controlled.

Analytical Variable.

The temperature.

Detection Systems.

Thermometer to measure temperature.
The operator's eye to observe the melting-point.

Sample Preparation.

The dry powdered sample is put into a small tube for insertion into the heating block.

Interpretation.

The interpretation of the results is carried out externally by the operator.

The examples cited above are very simple and follow the diagrammatic approach given in Fig. 1.4a. However, with the very wide range of analytical instruments which employ many different types of physical processes, we shall find that the simplified Fig. 1.4a does not cover all the possibilities.

1.4.5. Identification of Errors

It is useful to use the functional approach to an analytical instrument if you wish to try to assess where errors may be occurring in your analysis. In this way it is clear that you must consider such factors as the following.

— Have you chosen the correct experimental process for the particular analysis you require?

— Is the instrument measuring the variable you want?

— What sort of errors are present in the measuring system of the instrument?

— Can you (or the instrument) control the 'experiment' accurately enough?

— Is your sample in the correct form? Have you performed any sample pretreatment (such as dilution) accurately?

— Is you interpretation of the data correct?

SAQ 1.4a	By using the principal functional elements of an analytical instrument as headings in your answer, describe in note form the way in which a colorimeter works.

SAQ 1.4b   In what ways could an operator cause errors in measurement when using a melting-point apparatus?

An ingenious student is probably capable of finding many ways of misusing the instrument. However, for the purpose of this question consider the most probable errors, and associate these with particular functional elements given in Section 1.4.4.

1.5. CLASSIFICATION OF CHEMICAL INSTRUMENTATION

1.5.1. Introduction

It is not possible to classify rigorously all the different types of instruments that a chemist may use to perform an analysis. However, it is necessary to be aware of certain broad categories of instrumental methods, because the very act of classification has historically defined different branches of chemistry – eg spectrophotometry, chromatography.

By using the work we have done in these first few sections we can identify three main types of instrument:

— a Simple Measuring Instrument,

— a Measuring Instrument with a Selective Transducer,

— an Analytical Instrument with its own Experimental Procedure.

Further classification under those headings is done by looking at the experimental process which is being used.

1.5.2. Classification of Instrumental Analytical Methods

There is a great variety of experimental processes used in analytical instruments. However, the most common of these methods can be grouped into the four sections given below. In each section some examples are given of possible experimental processes, together with typical analytical variables. This classification is intended only to give you an impression of the breadth of instrumental methods available to you as an analyst. Explanations of the various topics will be developed in other ACOL Units.

Spectroscopy

Spectroscopy covers a very wide range of instrumental methods. The essential features of all these methods are:

— that they rely on the interaction of electromagnetic radiation with the molecules, atoms, or nuclei of the sample,

— that an important factor in the measurement, is the way in which the magnitude of this interaction changes for different wavelengths (or frequencies) of radiation.

Spectroscopic methods are found for all regions of the electromagnetic spectrum from radio waves to gamma rays. They use many different types of interactions (eg emission, absorption, scattering, diffraction).

There are other methods which also employ an interaction of radiation with matter but, because a change of wavelength is not a critical factor in the instrument, they are not called spectroscopic methods. An example of a non-spectroscopic optical method is use of the *refractometer*, which measures the refractive index of a sample.

Thermal Methods

Thermal methods all involve the measurement of some property of a sample that may change its value when the temperature of the sample is either increased or decreased. If, in thermogravimetric analysis for example, a sample is slowly heated, then its weight may decrease at certain specific temperatures because gas is given off when the sample changes its composition at those temperatures. An accurate record of the temperature at which the weight changes occur, will help to identify the particular constituents of the sample.

There are also other properties that may change with temperature, but it is unnecessary to list them now. You will be introduced to the range of possible methods in later units.

Electrochemical Methods

This group is probably the least coherent of the four categories of experimental methods that we are giving as examples. In general, electrochemistry refers to the relationship between an electric current or voltage and the behaviour of a sample in solution.

An electrochemical cell may generate its own voltage, and a measurement of the magnitude of that voltage is then of value in analysing the sample. Alternatively, we may apply a voltage or current to the sample in solution and observe the result. Other electrochemical methods may be employed as an aid to a wider analytical experiment, either as a means of detecting the end-point of a titration, or as a means of physically separating the constituents of the sample by electrolysis.

The common element is this category is the use of the interaction between electricity and a sample in solution.

Chromatographic Methods

Although there are many different kinds of experimental method in chromatography, these methods all depend on a basically similar concept. There are two phases: a stationary phase and a mobile phase. Depending on the particular type of chromatography, the mobile phase may be either gas or liquid, and the stationary phase may be liquid or solid.

In this group of methods, an actual physical separation of the various components of an initial mixture (the sample) can be achieved. The sample is carried through the system in the mobile phase, and the rate at which a particular constituent travels depends on the extent of its relative association with the stationary phase and with the mobile phase. Different constituents which exhibit a different preference, will travel at different speeds and hence arrive at the detector at different times. The separation occurs both in real time and as a physical separation of constituents.

We have seen that some types of experimental processes are so common that the particular group of methods is given a special name (eg chromatography). This name then, actually defines a whole branch of instrumental analysis. However, there are still many instruments and methods that do not fall neatly into one of the four categories given above, and as there is no apparent limit to the ingenuity of those devising new instrumental techniques, you are well advised always to be ready to meet new instrumental methods. The next paragraph may give you some help when you do meet an unfamiliar method.

1.5.3. On Meeting New Instrumental Methods

There is very little point at present in trying to present a complete table of the different types of experimental processes and their associated analytical variables. Many of them might be totally unfamiliar to you, as new techniques are constantly being developed. However, as you progress through this course, and indeed after you finish the course, you will meet many new methods and it would be useful, each time, to ask yourself the following questions.

What experimental process is being used?

What analytical variables are being measured?

What are the control functions? How can the operator change the conditions under which the 'experiment' is being performed?

When you feel that you know how the instrument works, then you should find out both the *capabilities* of the instrument, and its *limitations*. By asking these questions, and finding the correct answers, you can check whether you understand the principles of the particular instrument.

SAQ 1.5a   The term 'chromatography' covers a wide range of analytical procedures and instruments. Which of the following statements most clearly explains why it is possible to use the one name to cover all of the instruments?

(*i*)   The measurements can be made only on coloured solutions.

(*ii*)   They all employ broadly similar methods for measuring the amount of each component once they have been separated in the instrument.

(*iii*) They all use broadly similar methods to separate the different components of the sample.

(*iv*)   The instrumental methods all relate to a limited range of chemical compounds.

SAQ 1.5b   What questions should you ask yourself when you are confronted with a new and unfamiliar analytical instrument that you wish to use?

Summary

In the introduction the way in which we might begin to identify different categories of chemical instruments is discussed. We use the concept of *functional elements* which divides the instrument into separate units, each of which performs a particular function. Signals of both electronic and non-electronic form can then convey information through the instrument.

The idea of a simple measuring instrument is developed first, and this is then expanded by considering the operation of some sophisticated transducers (eg a pH-electrode). A further expansion of the simple measuring instrument leads to the concept of the analytical instrument as an instrument within which an actual experiment is performed. Finally the various branches of instrumental analysis are identified according to the types of experimental process used.

Objectives

It is expected that, on completion of this Part, the student will be able to:

- use the 'functional element' concept to describe a simple measuring instrument,

- predict the operational behaviour of some transducer systems under given conditions,

- use the concept of a 'signal' when describing the workings of an instrument,

- describe the operation of an analytical instrument in terms of 'experiment, measurement, and interpretation'.

2. Electricity in Instruments

Overview

The possible roles played by electricity in an instrument (carrying signals and supplying power), and the basic techniques of electrical measurement, are introduced at an early stage to reduce some of the mystery surrounding the electronics aspect of instruments. This is preceeded by a review of some of the basic principles, presented with their relevance to instrumentation.

2.1. REVIEW OF SOME BASIC PRINCIPLES OF ELECTRICITY

2.1.1. Charge, Q

The basis of all electricity and electronics is *electric charge*, Q. You should already be familiar with the electric charge of particles within the atom. The neutron has no charge, but the proton has a *positive* electric charge, and the electron has a *negative* electric charge. The charge on the proton is $+1.60 \times 10^{-19}$ coulombs. The charge on the electron has the same magnitude, but it is negative, *viz* -1.60×10^{-19} coulombs.

Electric charge is measured in *coulombs* (abbreviated C). This is a very large unit when we use it for measuring the static charge on an electron, but it is convenient when many electrons *move* giving an electric *current*. An even larger unit is the *Faraday*, which is equal to 96 485 C.

∏   Do you know the significance of the Faraday? Try the fol-
    lowing.

    (*i*)   Calculate the number of electrons that would be re-
            quired to give a total charge equal to the Faraday.

    (*ii*) Look up the Avogadro constant in Table 4 at the end
            of this Unit.

    (*iii*) What is the relationship between the two numbers?

(*i*)   1 Faraday is equivalent to 96 485 C. The charge on the elec-
        tron is -1.60×10^{-19} C. Therefore, the number of electrons
        required $= 96485/1.60 \times 10^{-19}$

$$= 6.03 \times 10^{23}.$$

(*ii*) The value of the Avogadro constant $= 6.02 \times 10^{23}$

(*iii*) The numbers are identical (within the accuracy of the figure
        given for the charge on the electron). The Faraday is the total
        charge (apart from the negative sign) corresponding to *1 mole*
        of electrons.

2.1.2.   Current, *I*

When an electric charge moves then there is an electric *current*.

Electrons in a metal wire are in constant random motion, but without an applied potential difference there is no *net* motion in any
direction. However, when a potential difference (and hence an electric field) is applied to the wire the electrons drift in one direction,
resulting in an electric current flowing along the wire – Fig. 2.1a.

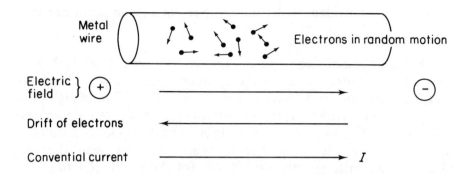

Fig. 2.1a. *Electron-flow and electric current*

The reason that electron-flow and electric current are in opposite directions is that the phenomena of electricity were discovered before it was known that it was mainly electrons that carried 'electricity' in metals. The result of this is that the 'direction' of an electric current has been *defined* as the direction of flow of a *positive* charge: it was not known that the current was due to the flow of negatively charged electrons. This should not cause any problem – it is merely a matter of convention. Throughout this unit we shall continue to use the direction of an electric current as being the direction of flow of *positive* charges.

The magnitude of the current, *I* is given by the number of coulombs of charge passing a fixed point every second.

$$I = dQ/dt \qquad (t = \text{time in seconds})$$

1 Ampere (A) = rate of flow of charge when 1 coulomb passes a fixed point in one second. You can see now that the coulomb is a convenient unit for charge when its motion is causing an electric current!

∏   We have not finished with big numbers yet! If a current of 1.00 A is flowing along a wire, how many electrons *per second* are moving past a fixed point on the wire?

1 ampere is equivalent to 1 coulomb passing a given point per second.

To obtain 1 coulomb we require $1/1.60 \times 10^{-19} = 6.25 \times 10^{18}$ electrons. This is indeed a very large number, and it may be imagined that the electrons would have to travel very fast so that enough pass the given point. However, the density of electrons (the number per unit volume) free to carry the electric current in copper is extremely high, 8.47×10^{28} electrons per cubic metre, and this means that the electrons are drifting only *very slowly* (of the order of 1 mm per second) even for quite high currents.

2.1.3.   Electric Potential, V

The electric potential at a point is equivalent to an *electric pressure*.

If we wish to see whether a current will flow between two points a and b, then we must ask if there is a difference between the *electric potential* at a and at b.

$$V_{ab} = V_a - V_b$$

V_{ab} = *Potential difference* between a and b.

V_a = *Potential* at a

V_b = *Potential* at b

Obviously, $V_{ba} = -V_{ab}$

If there is a *potential difference* ('pressure' difference) between a and b, and if we connect a conductor between a and b, then a current will flow.

It is very easy to confuse *potential* with *potential difference*. In fact, both are measured in *volts*, and sometimes referred to as *voltages*. Both are often represented by the same symbol, V. The *absolute* value of the *potential at a point* is taken to be the difference between the potential at that point and that at some arbitrary reference point, which has been chosen as the point of zero potential. The actual zero reference point is chosen to suit the particular problem, although it is frequently taken to be the 'Earth' connection in the instrument designated by the symbol shown in Fig. 2.1b.

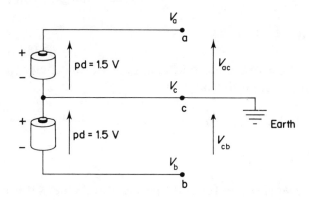

Fig. 2.1b. *Potential and potential difference*

For example, in Fig. 2.1b, two 1.5 V batteries are connected in series, and their common connection is connected to 'Earth'. Then

$$V_{ac} = +1.5 \text{ V}$$

$$V_{cb} = +1.5 \text{ V}$$

Since 'c' is connected to 'Earth', we have

$$V_c = 0$$

then,

$$V_a = V_{ac} = \qquad +1.5 \text{ V}$$

$$V_b = V_{bc} = -V_{cb} = -1.5 \text{ V}$$

$$V_{ab} = V_a - V_b = \qquad +3.0 \text{ V}$$

The potential difference (pd) between two points in a circuit is *defined* as being equal to 1 volt if 1 joule of electrical energy is changed into some other form of energy (eg heat or light) when 1 coulomb of electricity passes.

$$V = W/Q \qquad\qquad W = \text{energy converted}$$
$$Q = \text{charge flowing}$$

$$W = V \times Q$$

W is also the *work done* by the source of electricity in driving the charge, Q, through the pd V, volts

If we divide both sides by the *time* taken for this, t seconds,

$$W/t = VQ/t$$

W/t is the work done (energy) per second. This is the *Power*, P, in watts.

Q/t is the current flowing, I A.

Thus,

Electrical power, $P = V \times I$ (volts $\times$ amps $=$ watts)

This is important because it shows us that to obtain *power* from an electric circuit, we need *both* voltage (pd) and current. For example, a mains outlet socket may give a pd ('pressure') of 240 V, but it will not supply any power until something, which draws a current, is plugged into the socket.

2.1.4.   Relationship Between Voltage, V and Current, I

The way in which the current I might change for different values of applied pd, V, depends on the particular circuit or device that is connected to the source of pd.

The simplest case is that of a resistor, magnitude $R\Omega$ [Ω = ohm(s)]. If we refer to Fig. 2.1c, the effect of an applied pd, V_{ab}, will be the flow of a current I_{ab}.

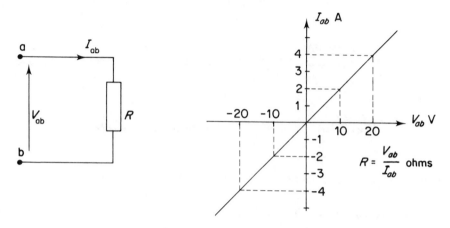

Fig. 2.1c. *Ohmic behaviour*

Doubling the value of V_{ab} will result in a doubling of I_{ab} (and so on), giving a *straight-line graph*. Note that if the pd is reversed (+ to −) then the current also changes direction (+ to −), but the values can still be plotted on the same graph. The plot of I_{ab} versus V_{ab} can be called the 'characteristics' of the resistor. For a given pd, V_{ab}, we can calculate the current I_{ab}.

Note that the ratio V_{ab}/I_{ab} is the same for all points on the graph. This constant ratio is called the *resistance*, R, of the resistor, and the calculation leads to Ohm's law:

$$V = I \times R$$

Note that the subscripts, a and b, have now been dropped, however 'I' is still the current *through* the resistor and V is the potential *difference* between the ends of the resistor.

∏ Calculate the current through the two resistors shown in Fig. 2.1d.

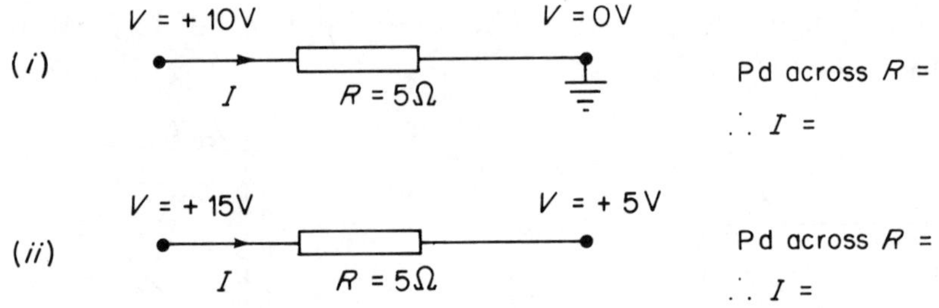

Fig. 2.1d.   *Ohm's law and potential difference*

(*i*)   The pd, V, across the resistor $= 10 - 0 = 10$ V.

Therefore $I = 10/5 = 2$ A.

(*ii*)   In this case the potentials at the ends of the resistor are 15 and 5 V respectively, therefore,

pd, V, $= 15 - 5 = 10$ V

Therefore, $I = 10/5 = 2$ A.

In applying Ohm's law you must be careful to remember that V refers to the potential *difference* between the ends of the resistance R, and not just the potential applied *at one end*.

The characteristics in Fig. 2.1c are a linear* graph – these are also called 'Ohmic' characteristics.

There are, however, many examples of non-Ohmic behaviour. One of these is the *diode* (see Fig. 2.1e) which will 'conduct' in only one direction. This means that a large current will flow only if the 'anode' of the diode is connected to the more positive potential. If it is connected the other way around, there will be no (or very little) current.

(* More correctly, rectilinear, but linear is commonly used.)

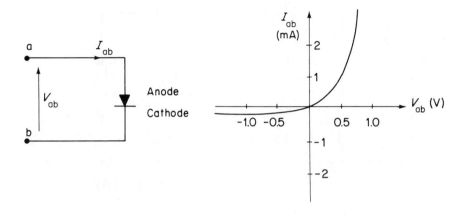

Fig. 2.1e.   *Non-Ohmic behaviour of a diode*

2.1.5.   Power in a Resistor

We have seen already that power, P, in an electric circuit is given by

$$P = V \times I,$$

and that this gives the rate at which energy is being converted into some other form. For a simple resistor the energy is converted directly into heat, and the rate at which the heat energy is produced can be calculated from the equation below.

$$P = V \times I = (I \times R) \times I = I^2 R = V^2/R$$

This is sometimes referred to as Joule heating.

2.1.6.   Potential Divider

One of the most important (and simple) circuits in electronics is the *potential divider*. As its name suggests this circuit, Fig. 2.1f(i), divides an initial (input) potential difference, V_i, giving an output potential difference, V_o, which is a *fraction* of the input voltage.

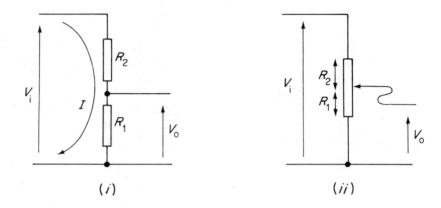

Fig. 2.1f.   *Potential divider*

We can calculate the current through the resistors, R_1 and R_2.

$$V_i = I \times (R_1 + R_2)$$

$$V_o = I \times R_1 = \{V_i/(R_1 + R_2)\} \times R_1$$

$$\therefore \qquad V_o = V_i \times R_1/(R_1 + R_2)$$

Thus the output pd (or voltage), V_o, is a fraction of the input pd (or voltage), V_i, depending on a ratio determined by the magnitude of the resistors. The ratio $R_1/(R_1 + R_2)$ may have any value from 0 to 1, thus V_o may have any value from 0 to V_i. It is frequently preferable that, instead of two separate resistors, we use a single resistance *track*, and move a contact along this track, as in Fig. 2.1f(*ii*). In this case $(R_1 + R_2)$ has a fixed value, but the value of R_1 changes as the contact slides along the track. As an example, this type of control is frequently used to adjust the volume in domestic audio-equipment by altering the signal amplitude, V_o. The track may be circular (for rotary control knobs) or linear as for the modern linear volume controls.

The general name given for this latter type of variable potential control is a *potentiometer* or just a '*pot*'.

2.1.7. EMF of a Battery

EMF is an abbreviation which is frequently used. It stands for *Electro-Motive Force*. We can understand it best by looking at one form of electrochemical cell – a 'battery' as in Fig. 2.1g.

We can measure the pd between the terminals of the 1.5 V torch battery, Fig. 2.1g(i), but what exactly shall we measure?

The actual source of *electrical pressure* arises from a chemical reaction inside the cell, Fig. 2.1g(ii), and it is this *pressure* that is referred to as the *electro-motive force* (emf). When we want to use this *emf* to do electrical work, we must complete the cell by making electrical connections to *terminals* which can be linked to the outside world.

In Fig. 2.1g(iii), we have drawn an electric circuit to represent the action of the battery. This *equivalent circuit* is intended to perform in the same way as the real cell would do when connected to the same external circuit. We can represent the 'emf' by the pd E. However, because the chemical solution, together with the connectors and terminals, have an electrical resistance we must also include an 'effective' resistance, R_0. We are unable to measure E directly, since we can connect only to the physical terminals. We can only measure the external pd, V volts. The term 'EMF' is thus normally found to refer to the actual 'source' of electrical pressure, and it is not always easy or even possible to measure it directly.

∏   For a battery with an emf, $E = 1.5$ V, and $R_0 = 1.0$ Ω, calculate the measured pd, V, between the two terminals, if the battery is connected to an external resistance of 2.0 Ω. See Fig. 2.1h.

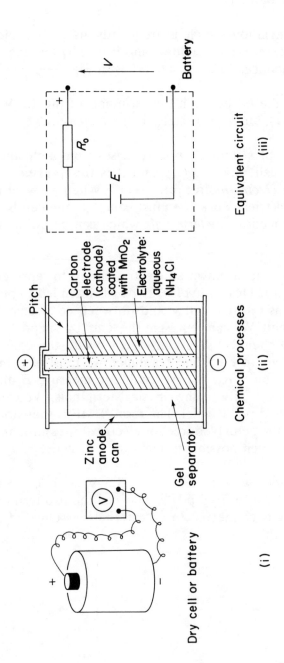

Fig. 2.1g. *Representations of a dry cell*

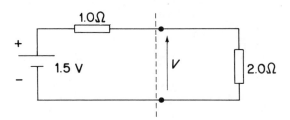

Fig. 2.1h. *EMF and voltage*

The total resistance in the circuit is 3.0 Ω. Thus the current following through the circuit is 1.5/3.0 = 0.50 A.

Thus the pd across the external 2.0 Ω resistance is 0.50 × 2.0 = 1.0 V. This is the voltage that appears at the terminals of the battery.

Another way of looking at the problem is that the current of 0.50 A causes a voltage drop of (1.0 × 0.50) = 0.50 V across the internal resistance, giving an output voltage,

$$V = E - I \times R_0 = 1.5 - 0.5 = 1.0 \text{ V}.$$

2.1.8. Variation of Voltage and Current with Time

An important characteristic of an electrical circuit is the way in which V and/or I change with time. Consider the simple example in Fig. 2.1i(i).

When the switch S is closed, a pd of 1.5 V is connected to the lamp and a current of 1.0 A flows. When the switch S is open, both V and I are zero. If the switch is opened and closed repeatedly for a period of one second each, the resulting variations of V and I could be recorded as in Fig. 2.1i(ii). This is in fact an example of intermittent DC. DC stands for *Direct Current*, which means that the current flows in *only* one direction (even if the current is sometimes zero).

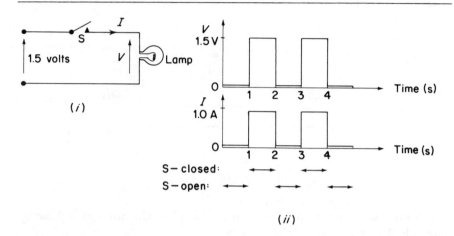

(i)

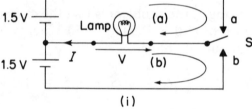

(ii)

Fig. 2.1i. *Intermittent direct current*

For an example of *Alternating Current*, AC, see the circuit in Fig. 2.1j(*i*).

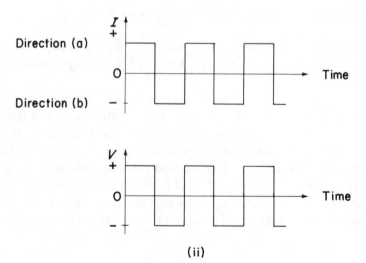

(i)

(ii)

Fig. 2.1j. *Alternating current*

In this case the switch S is alternately switched from contact 'a' to 'b'. When S is switched to a, the current follows the arrow, (*a*), and the pd, *V*, is positive. When S is switched to b, the current follows the arrow, (*b*), and the pd, *V*, is negative. The resultant time variation graph for *V* and *I* is given in Fig. 2.1j(*ii*) This is an example of a square wave.

2.1.9. Sinusoidal AC

An example is given above where an AC signal appears in the form of a square wave. However, the normal starting point for considering AC is the *sinusoidal wave*, as in Fig. 2.1k.

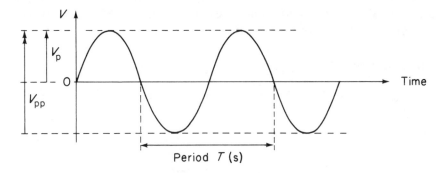

Fig. 2.1k. *Sinusoidal AC*

The variation of voltage (or it could be current) follows a sine wave symmetrically placed about the zero value. The amplitude of the wave from a negative peak to a positive peak is called the *Peak-to-peak Voltage*, V_{pp}. The amplitude from the 'zero' line to one peak (positive or negative) is called the *Peak Voltage*, V_p, and clearly,

$$V_p = 0.5 \times V_{pp}$$

However, neither of these values (V_p or V_{pp}) gives an ideal measure of the magnitude of the wave voltage, because the voltage only reaches these values for an instant before returning to some smaller value. What other value could be chosen?

If we take an *average* value of V ($\overline{V}$) the result will be zero ($\overline{V} = 0$), because there are equal amounts of positive and negative signal in the wave. If we take an *average* value of V^2 ($\overline{V^2}$), then this will not be zero (V^2 is always positive). However, ($\overline{V^2}$) is a voltage squared, and hence if we are to express the result simply in terms of a voltage, we must take the square root of this average.

Thus:

$$V_{rms} = \sqrt{\overline{V^2}}$$

(rms stand for Root-Mean-Square).

The particular advantage in choosing V_{rms} is that it is the voltage of an equivalent *DC signal* that gives the *same power* as the average power available in the *AC signal*.

For a *sine wave*,

$$V_{rms} = V_p/\sqrt{2}$$

or

$$V_p \approx 1.4 \; V_{rms}.$$

∏   If the mains voltage is given as 240 V, calculate the peak voltage.

The value of 240 V given for mains voltage can be used directly for calculation of power, eg in electric heaters. Hence 240 V corresponds to the Root-Mean-Square Value, V_{rms}.

Then $V_p \approx 1.4 \times V_{rms} \approx 1.4 \times 240 \approx 340$ V.

Looking again at Fig. 2.1k, the period T (in seconds) of the wave is given by the time taken for one complete cycle of the wave (from one point on the wave through both a positive and a negative peak back to the equivalent point on the next wave). The frequency, f, of the wave is the number of complete cycles per second, ie

$$f = 1/T$$

This is normally expressed in hertz (Hz), where 1 Hz is equal to one cycle per second.

2.1.10. Electric Mains Supply

In the UK the mains supply is normally about 240 V AC with a frequency of 50 Hz. Alternative supplies work at 110 V with a frequency of 60 Hz (USA system). Many systems (including that in the UK) use 3-wire connections to the instruments. The three connections are given below (using wiring colours appropriate to UK) – see Fig. 2.11.

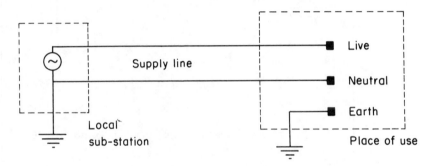

Fig. 2.11. *Mains supply*

Earth: Green or Green/Yellow wiring. This is connected directly to the ground at the building in which the supply is used. For example this connection may be made to a metal pipe carrying the mains water-supply. The electric potential is zero.

Neutral: Blue wiring (previously black). This is connected to 'ground' at the point from which the electricity has been supplied (eg a local substation), but it is *not* connected to 'ground' at the point of use. The electric potential will obviously be close to zero, but there may be a few volts present due to large currents flowing along the neutral wire between the points of supply and use.

Live: Brown wiring (previously red). This is the connection that carries the electrical pressure. For a 240 V mains supply the potential on this wire will vary sinusoidally between $+340$ V and -340 V.

SAQ 2.1a   Assume that the batteries in Fig. 2.1m have no internal resistance of their own, and calculate the current that flows through the 100 Ω resistor.

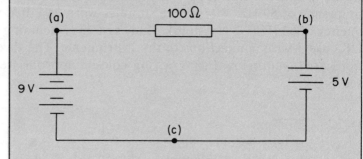

Fig. 2.1m

Which way will the current flow through the resistor,

(i) from 'a' to 'b', or

(ii) from 'b' to 'a' ?

SAQ 2.1b

A rotary 'pot' as in Fig. 2.1n has a circular resistance track of uniform resistance forming an arc of total angle 270 degrees. A wiper makes electrical contact to the track and can be turned so that the angle θ varies between 0 and 270 degrees.

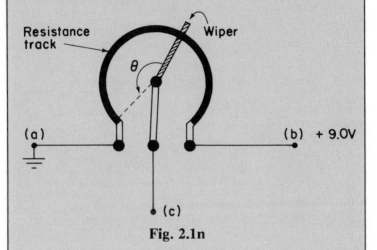

Fig. 2.1n

The terminal 'a' is connected directly to 'earth' and terminal 'b' connected to a potential of $+9.0$ V.

Calculate the angle θ required for the wiper if the centre terminal, 'c', is to have a potential of 3.0 V.

SAQ 2.1c   Calculate the AC current required for a 3 kW
                      water-bath heater to be run from the 240 V
                      mains supply.

                      State whether your value of the current is the
                      'peak' value or the 'root-mean-square' value.

2.2.   ELECTRICITY – SUPPLIER OF POWER OR CARRIER
         OF INFORMATION?

2.2.1.   Role of Electricity within an Instrument

It is important to realise that electricity may have two distinct functions within an instrument:

— to supply power ie to be a source of energy which can be converted into other forms – heat, light, sound, motion,

— to carry information – to be a convenient method of transferring, storing, or processing information in the form of electronic signals.

∏   Consider the following examples of the use of electricity.

(*i*)   Mains electricity may be used to heat an oven.

(*ii*)   A battery is used to 'light' a torch bulb.

(*iii*) A battery is used to drive a toy car backwards and forwards.

(*iv*)   An electric signal occurs when a push-button switch on a calculator is pressed. This signal 'tells' the calculator which number has been selected.

(*v*)   Mains electricity drives a synchronised motor. The speed of this motor is synchronised to the frequency (50 Hz) of the electricity supply.

(*vi*)   A pH-electrode produces a voltage which the pH-meter 'interprets' as the value of the pH of the solution.

The above examples can be grouped into three categories.

(*A*)   Where the electricity *only* carries the *power* (or *energy*) to make something happen. This electrical energy is converted into other forms of energy: motion, heat, light, sound, etc.

(*B*)   Where the electricity carries information but does not have enough *power* to make things happen by itself.

(*C*)   Where there is a mixture of both power and information.

List the above examples, (*i*) to (*vi*) in the categories (*A*), (*B*), and (*C*) in the table below. You may well be unfamiliar with the concept of *information* in this context and find some difficulties with this exercise, so do not spend too long on the problem.

(*A*):

(*B*):

(*C*):

The answer to the question is as follows:

(*A*):   (*i*) (*ii*)

(*B*):   (*iv*) (*vi*)

(*C*):   (*iii*) (*v*)

The mains supply and the battery in both (*i*) and (*ii*) are only supplying energy (or power) – category (*A*). The only real information involved is whether the oven and the bulb is 'on' or 'off'. *If* the torch were used for *signalling* (eg by switching on and off), then (*ii*) would go into category (*C*). (N.B. A change of mains frequency (eg 50 Hz to 60 Hz) would not affect the heating in the oven, nor would a change of polarity of the battery affect the light intensity from the bulb).

In a synchronised motor, the *speed* is controlled by the *frequency* of the AC signal. A motor designed to run in the UK on 50 Hz, AC, will run *fast* in the USA on 60 Hz. In this case the *frequency* carries information about the passing of time. The electricity also supplies the *power* to drive the motor, hence (*v*) goes into category (*C*).

With the toy car there is information as well as power carried in the electrical signal. If the battery is reversed the car will go *backwards*, the *sense* of the battery 'tells' the car which way to go – category (*C*).

Both the switch, in (*iv*), and the pH electrode in (*vi*) are signalling information – category (*B*). For the switch the choice between 'on' and 'off' is the extent of the information; this is, in effect, a *digital* signal. The pH-electrode, on the other hand, will produce an infinite number of possible values of voltage (within a maximum range of about 1 V, corresponding to the range of possible pH values). This type of continuously variable voltage is called an *analogue* signal.

2.2.2. Electric Power

We know from Section 2.1.3 that the electric power, P, is given by:

$$P = V \times I,$$

where V and I are voltage and current respectively. Thus, to supply power, it is necessary to supply both voltage and current. The requirements for voltage, current, and hence power, differ considerably amongst the different component parts of instruments.

Modern electronic circuitry requires very little power. It is possible for some instruments (eg small pH meters) to be driven for long periods from ordinary dry batteries. However, there are still several components within chemical instrumentation that will always require a significant amount of power. These include electric motors, heaters, and sources of radiation. By far the most common primary source of power is the normal electric mains supply (*ca.* 240 V in UK). This can either be used directly as AC Mains Power, or it can be transformed into some other electrical source, usually DC. We discuss some of the important aspects of these two types of supply in the following sections.

2.2.3. AC Mains Power

The mains supply is used directly to provide the power for most *heaters*, eg in a bench heater-stirrer mantle. Normally the rate of supply of energy has to be controlled, and in most applications this is done by *switching* the current on and off for different periods, depending on the amount of power required. With a heating mantle, a *simmerstat* is often used. This is a simple switching circuit housed in a self-contained unit. When full power is required the switch is 'on' all the time. For half-power the switch is 'on' and 'off' for similar periods of about 2 seconds each, and so on. Other power levels are obtained by altering the ratio between the 'on' and 'off' times. The same system is used in the normal domestic iron. The periods during which the heater is actually 'on' can sometimes be seen by means of a small indicator light, which then glows.

It is important to realise that if a 'simmerstat' system is used, it is only the amount of electrical *power* that is controlled. This does not necessarily mean that the *temperature* is being kept at some fixed value. The actual temperature reached depends on how quickly the heat is being conducted away from the heater – eg a bench heating mantle on full power might reach a temperature of 250 °C without anything on it, but, with a large beaker of boiling water to conduct away the heat, it will reach only just over 100 °C.

For more sophisticated control of power, modern circuits use a new semiconductor device called a TRIAC, which is a very fast solid-state switching device. Instead of switching 'on' and 'off' in a matter of seconds, as with the simmerstat, the triac can switch on and off in less than one millisecond. It is so fast that it can be used to control the power to a normal lamp without causing any apparent flicker in the light output. The triac is used in domestic light-dimmer circuits. It can also be used as a very simple control for the speed of small motors – eg to control the stirring speed in some heater-stirrers.

∏   A TRIAC can be used in a simple control circuit to control the amount of *power* going to a water bath in the same way that a TRIAC can be used in a lamp dimmer-control to control the brightness of a light. However, keeping the power input constant is not sufficient to ensure that the *temperature* of the bath remains constant.

Describe, by using a very simple block diagram, an instrumentation system that could be used to keep the temperature of a system, such as a water bath, constant at a temperature selected by the user.

Hint. You will have to use some form of measurement transducer.

The 'measurement transducer' that must be used is obviously some form of temperature transducer (see Section 1.2.7). A thermistor (see Section 1.3.3) could be used, although there are several other possibilities: eg a bimetallic strip, or an expansion thermometer. The output from the transducer would have to be fed *via* some form of signal processor into a *control unit* which would actually cause the TRIAC to switch 'on' and 'off' in the correct way. There would also have to be another input into the control unit to 'tell' the unit what temperature had been selected. This would normally take the form of a simple temperature-calibrated rotary control.

The type of system just described, uses a process known as *negative feedback*. This is where information from the *output* of the system is *fed back* into the *input* of the system in such a way as to make the output more *stable*. In this example, the temperature of the water bath is measured, and this information is used to adjust the system to bring the temperature to its correct value. *Positive* feedback acts in the opposite way and tends to make a system more unstable, and with positive feedback, it may actually start oscillating from one extreme to the other. You will come across many examples of *feedback* in very many different forms.

2.2.4. Transformation of Mains AC into DC

Very many components and circuits within an instrument require a DC electrical power supply. This is mostly at a low voltage (between about 5 and 30 V), although some components need higher voltages, eg photomultipliers (*ca.* 1000 V), discharge lamps (*ca.* 100 V).

A simple circuit for obtaining low voltage DC from the AC mains is given in Fig. 2.2a, and we discuss the various parts of that circuit in the following paragraphs.

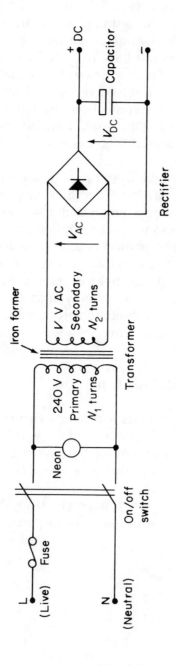

Fig. 2.2a. *DC voltage from AC mains supply*

A transformer is constructed of at least two sets of windings (primary and secondary) wound on an iron former, and is used to convert the AC mains supply into an AC supply of a lower voltage. If the input is at 240 V, then the output AC-voltage, V volts, is given by

$$V = 240 \, N_2/N_1$$

where N_1 and N_2 are the number of turns on the input (primary) and output (secondary) windings respectively. Thus the output from the transformer would be given by Fig. 2.2b(i). This AC signal then enters the *rectifying* circuit. This little circuit contains diodes (see Section 2.1.4) which allow the current to pass in only one direction. This has the effect of converting an AC signal into a DC (single direction) signal, as in Fig.2.2b(ii).

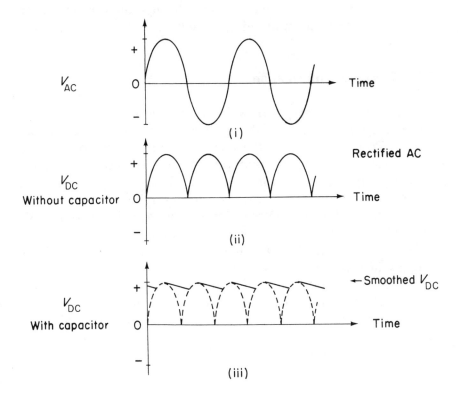

Fig. 2.2b. *Rectification and smoothing*

Although the signal in Fig. 2.2.b(*ii*) is DC (because it flows only in one direction) it is still not very useful as a power source because the magnitude is continually varying between maximum and zero.

The next step is to use some method of making the magnitude of the voltage more uniform. This is done by using a *capacitor*. A capacitor is a device which is capable of storing *charge*, and because it acts as a reservoir of electric charge, it helps to maintain the electrical pressure (volts) at a nearly constant level, see Fig. 2.2b(*iii*). A final smoothing of the voltage can be achieved, if required, by using an electronic circuit for voltage stabilisation.

∏   As an exercise you should look at a circuit diagram given in the manual of some analytical instrument and try to find the mains input on the diagram, with the associated fuse and on/off switch, and then trace this through the transformer to the rectifier and smoothing capacitor.

2.2.5. Electrical Information

The sections above have dealt with the supply of electric *power* to instruments for heating or lighting or as the driving force for machinery. We now introduce the idea of electricity also being used as a means of carrying *information*. This information may be exceedingly *complex* as for the electronic signal being fed into the aerial socket of a television receiver and carrying the information necessary to produce the colour pictures and associated sound. Alternatively, the information may be very *limited*, eg when the neon light on an oven control shows when the heater is switched on. There are many methods of conveying information with an electrical signal. All these methods require the signal to have some measurable quantity which can be used to represent a magnitude in the information. The various possibilities are discussed in the following paragraphs.

2.2.6. Ways of Conveying Information with DC Signals

The simplest and most common indicator of electrical information is a light on a control panel which may be either on or off. This may convey information that the instrument has or has not been switched on. To make the light operate requires both voltage and current. However, there are examples where it is only the voltage *or* the current that is important in carrying information.

A pH-electrode gives a *voltage* signal which is related to the pH of the solution. If a current in excess of a few nanoamps is taken from the electrode then the voltage is reduced from its correct value. Ideally, *no current at all* should be allowed to flow.

A photovoltaic cell will record the intensity of light falling on it. The magnitude of the *current* from the photo-cell is proportional to the intensity of the light.

In each of the above examples the information is carried in the *magnitude* of the signal (voltage or current).

There is one further way in which a DC signal can pass on information. It is possible to *reverse* the direction of the voltage or current, as for the toy electric car. Depending on the *polarity* of the battery, the car will move either forwards or backwards.

Thus we can see that a *DC electric signal* may convey information in any of the three forms:

— voltage,

— current,

— polarity.

2.2.7. Ways of Conveying Information with AC Signals

What about an AC signal as shown in Fig. 2.2b(i)? Obviously information can be carried by changing the magnitude of either *voltage* or *current* (as for DC).

However, it is impossible to assign a 'polarity' to a signal that is constantly changing from positive to negative. Polarity is therefore not a parameter that can be used for the AC signal. However, we see below that there are two other 'information parameters' for AC signals.

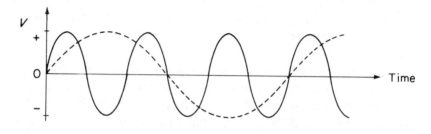

Fig. 2.2c.   *Waves with different frequencies (or wavelengths)*

In Fig. 2.2c, two signals have been drawn. They both have the same magnitude (amplitude), but they are clearly different. The difference is in their *frequency*, and if we have some device to distinguish between the different frequencies of the two waves, then we can use *frequency* as an information-carrying parameter. This same situation is already familiar to us in sound – our ear can easily differentiate between a note of 256 Hz (middle C) and one of 440 Hz (A above).

Not unexpectedly, electronic circuits *are* capable of distinguishing between electronic signals of different frequencies. Indeed 'frequency selective' circuits (or filters) can select a narrow range of frequencies from an extremely complex signal. One popular example of this differentiation is the 'disco' lights that are linked to the music – it is not uncommon for different coloured lights to have their brightness controlled by low, middle, and high frequency ranges in the music. High frequency (treble) notes may switch on the red lights, for example, and the low frequencies (bass) may switch on the green.

Now look at the Fig. 2.2d. Again there are two signals which are clearly different although they have the same amplitude and the same frequency.

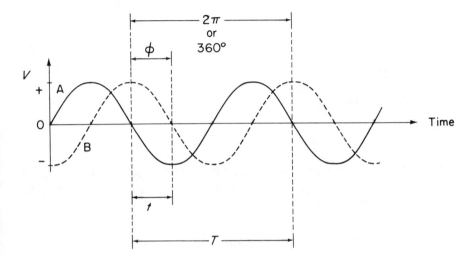

Fig. 2.2d. *Waves with different phases*

The difference in this case is one of *timing*. The positive maximum of the wave A occurs when wave B is at zero. If we relate the difference in time, *t*, to the period of the wave *T* then we can get a factor called the *phase difference*, θ, expressed either in radians or in degrees.

$$\theta = 2\pi t/T \text{ radians } = 360t/T \text{ degrees.}$$

In the example in Fig. 2.2d the phase difference, θ, is $\pi/2$ radians or 90 degrees.

The actual *phase* of an AC signal is a term which is used to describe the position of the wave in its cycle. The length of one cycle is the distance between successive points on the wave which have the *same* phase. The phase of an AC signal can be used to carry information. In order to interpret this information it is necessary to compare the AC signal with some *reference* wave of the same frequency and obtain a value for the phase *difference*, θ. The electronic circuit that 'compares' a reference wave with a signal wave may be called a 'phase sensitive detector' or in some cases a 'synchronous rectifier'.

Thus for an *AC signal* the information may be carried by the following parameters:

voltage,

current,

frequency,

phase.

∏   A phase-sensitive detector (PSD) or synchronous rectifier (see Fig. 2.2e) is a circuit which will produce a DC voltage, V_{dc}, which depends on:

— the magnitude of an input AC signal, V_{ac},

— the phase difference, θ, between the AC *signal* and a *reference* wave.

$$V_{dc} = K V_{ac} \cos \theta$$

where K is a constant which is the *gain* of the particular PSD, and V_{ac} is the peak value of the *signal* wave. Note that the magnitude of the DC output voltage does not depend on the magnitude of the *reference* wave.

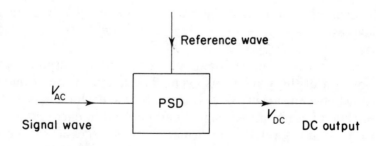

Fig. 2.2e.   *Phase-sensitive detector*

(*i*) If $K = 1$, and the peak value of the input signal is 10.0 V, calculate the DC output voltage for the following phase differences, θ, between the signal and the reference wave:

    0, 45, 90, 135, 180, 225, 270, 315, 360 degrees.

(*ii*) From your results decide whether or not any 'information' could be lost when an AC signal is converted into a DC signal *via* a PSD.

To answer this question merely requires the calculation of some cosine values.

The *importance* of the question however, lies in the fact that it shows that it is possible to use an appropriate circuit (a PSD) to convert information carried as the *phase* of an AC waveform into information carried by a combination of the *magnitude* and *polarity* of a DC signal.

(*i*) Phase differences/degrees:

    0, 45, 90, 135, 180, 225, 270, 315, 360

    V:

    $+10, +7.1, 0, -7.1, -10, -7.1, 0, +7.1, +10$

(*ii*) It should be clear that some information *is* lost on passing through the PSD. Note for example that a DC signal of magnitude $+7.1$ could correspond to a phase difference of either 45 or 315 degrees, with the result that the DC output cannot differentiate between the two different inputs. In most applications the system is designed so that this does not cause the loss of any important information!

2.2.8. Other Ways of Conveying Electrical Information

An alternative form of electrical signal is in Fig. 2.2f(*i*) and (*ii*). Note that the magnitude of the signal is given as 5 V in each case,

so that the voltage (or current) magnitude does *not* represent one
of the variable parameters. The difference between the two signals,
(*i*) and (*ii*), is in the sequencing of the pulses. Each signal has a
different *code*. Provided that we have electronic circuits that are
capable of interpreting the code, then we can use this as a method
of conveying information.

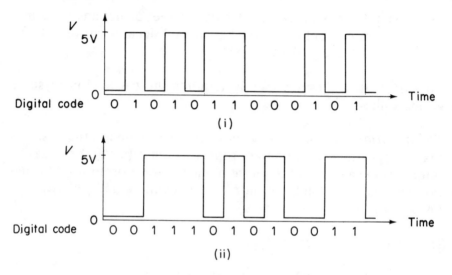

Fig. 2.2f.   *Digital coding of signals*

The code is represented below each signal in the diagram as a series
consisting of '0' and '1'. These are *digital* codes, and they are already
becoming familiar in shops as the black and white bar code that
may be used to identify many of the products on sale. A light-pen
is used to read this code and sends the information as sequence of
pulses (digital signal) into the computing system used in the cash
register. Thus a digital signal does not use voltage, current, polarity,
frequency, or phase (directly) as a means of conveying information,
but instead relies on a form of *code* that is incorporated into the
signal.

We shall see in later units how these information signals are used
in instruments, and also how it is often necessary to convert information from one form into another, as in *signal processing*.

SAQ 2.2a

Does the battery for a digital watch supply

(*i*)   power,

(*ii*) information,

(*iii*) a combination of both power and information?

Please mark one of these answers.

SAQ 2.2b

Indicate which parameter(s) below may convey information in the DC, the AC, and the digital mode.

1. Voltage. 2. Current.   3. Frequency.

4. Polarity. 5. Coded pulses. 6. Phase.

Write the appropriate numbers (1–6) alongside each of the following:

DC

AC

Digital

SAQ 2.2b

SAQ 2.2c   The radiation source in a certain simple spec-
             trophotometer is a 12 V (DC), 24 W, tungsten
             lamp. This is supplied by a circuit similar to that
             in Fig. 2.2a. Calculate the minimum current that
             must be drawn from the 240 V mains supply.

             Hint: You need be concerned only with the con-
             servation of energy.

SAQ 2.2d

Refer to Fig. 2.2g, and calculate:

(i)   the frequency of each wave,

(ii)   the phase difference between waves S and R,

(iii) the DC voltage that would be obtained from a PSD with $K = 1$, (see Section 2.2.7) if S is the signal wave and R is the reference wave.

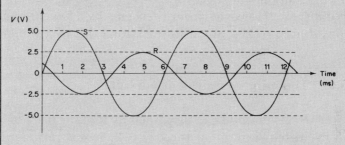

Fig. 2.2g.   *Wave forms for use in SAQ 2.2d*

2.3.   EQUIVALENT CIRCUITS AND ELECTRICAL
MEASUREMENTS

2.3.1.   Introduction

We have already come across several components which give an
electric signal: pH-electrode, dry-cell battery, thermal conductivity
detector, photocell, electrical mains supply. The means by which
the electric signal is generated is different in each case, and so too
are the characteristics of each signal.

This diversity might suggest that each would have to be treated in a
separate way. However, from theorems by Thévenin and Norton, it
is possible to treat many different situations in a similar way.

2.3.2.   Thévenin's Equivalent Circuit

A clue to this method can be gained by looking back to Section 2.1.7
and the discussion about the dry cell. It was proposed that within
the dry cell there is a source of electrical pressure, the emf, E, and
that associated with the chemical solution and the terminals there
is also an 'effective' resistance, R_0. E and R_0 cannot be separated to
get either the pure emf or the pure resistance of the battery, but we
can say that the battery behaves *as if* it was constructed of a source
of emf, E, and a resistance, R_0.

Thus we can regard the circuit in Fig. 2.3a as behaving in the same
way as the actual battery. This is the *equivalent circuit* for the battery. In fact we have already used this 'equivalent circuit' in Section
2.1.7 to calculate the change in the output voltage of a battery if a
large current is drawn from it.

One theorem in electricity that covers this topic is due to Thévenin.
It is unnecessary to study this in detail, except to know that his con-
clusion was that it is possible to use the *same equivalent circuit*, as
in Fig. 2.3a, for many *different sources* of electricity. It is necessary
only to specify the value of R_0 and the value of E.

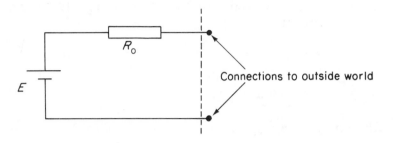

Fig. 2.3a. *Thévenin equivalent circuit for source of DC*

Let us take, for example, the pH-electrode. If we use the equivalent circuit in Fig. 2.3a, we know already from Section 1.3.2 that the emf in mV, E_V, is given by:

$$E_V = E_0 - 0.198 \times T \times (\text{pH} - 7.0)$$

However, we have not met anything yet which tells us the value of R_0. This turns out to be very high, of the order of millions of ohms (or mega-ohms, MΩ). The high resistance is due principally to the glass membrane – it is difficult for a large number of ions (carriers of electricity) to flow quickly through the glass.

∏   Consider a pH-electrode, and assume that the pH of the solution is such that the emf of the electrode is 300 mV, and its internal resistance is 1 MΩ. What would be the potential difference at the terminals of the electrode if it were connected to an external resistance of 10 MΩ?

By using a Thévenin equivalent circuit for the pH-electrode with E = 300 mV and R_0 = 1 MΩ, we obtain a complete circuit (with R_i = 10 MΩ) as in Fig. 2.3i.

The total resistance in the in the circuit is 11 MΩ, thus the current flowing will be given by:

$$I = 300 \times 10^{-3}/11.0 \times 10^6 = 2.73 \times 10^{-8} \text{ A}$$

The actual voltage measured is that which appears across the external resistor of 10 MΩ.

$$V = I \times R = 2.73 \times 10^{-8} \times 10 \times 10^6$$

$$= 273 \text{ mV}.$$

Thus there is a percentage error of $(300 - 273)/300 \times 100\% = 9\%$ in measuring the true emf.

We shall see later how this error in measurement can be minimised.

It is unlikely that an equivalent circuit would ever completely mirror the behaviour of the real thing. For example, the equivalent circuit of our battery does not show the phenomenon of 'polarisation' that occurs with a real battery if a very heavy current is drawn from it (in effect R_0 increases due to a disturbance of the relevant chemical equilibrium). However, within certain limitations, *equivalent circuits* are extremely useful.

2.3.3. Norton's Equivalent Circuit

We have mentioned in Section 1.3.1 that some transducers produce an *electric current* that is proportional to the input signal. It is then useful to call on the theorem by Norton, which states that we can use an *equivalent circuit* of the form given in Fig. 2.3b. The idea of a *current generator* instead of a voltage generator (or emf) may be new to you, but it is a very useful concept. We assume that the 'current generator' is constantly producing a current, I_0, and that this current can either flow entirely within the source (through R_0), or, if an external load R is connected, part of I_0 flows out of the source and through R.

As an example of a current-generating transducer, we can consider a simple photovoltaic cell in which the magnitude of the current generated, I_0, is proportional to the intensity, L, of the incident light.

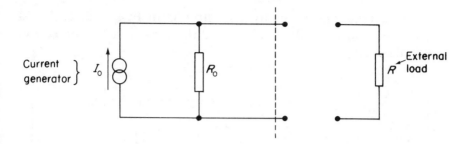

Fig. 2.3b. *Norton's equivalent circuit*

Here, however, the equivalent circuit, Fig. 2.3c, is a little more complicated than the basic Norton circuit in that, in addition to a resistance R_0, there is a diode which begins to conduct when the voltage, V, is more than about 0.3 V, and has the effect of limiting the maximum output voltage to about 0.5 V irrespective of the intensity of the light (hence voltage is not a good indication of a high light-intensity).

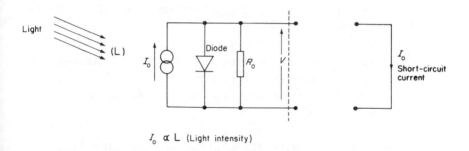

$I_0 \propto L$ (Light intensity)

Fig. 2.3c. *Equivalent circuit for photovoltaic cell*

If we short-circuit the output terminals of the cell (ie connect them directly together) with a piece of wire, then *all* the current I_0 will flow out of the cell, through the wire and back to the cell. If we can also measure the *magnitude* of that current I_0, then we can use that magnitude to give the intensity of the light, ie we have a simple light-meter.

∏   Consider the two equivalent circuits in Fig. 2.3d. One is a
      Thévenin equivalent circuit and the other is a Norton equiv-
      alent circuit.

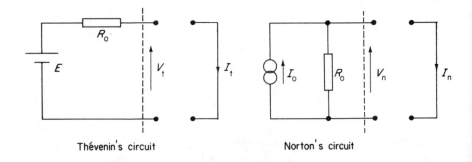

Thévenin's circuit   Norton's circuit

Fig. 2.3d.   *Equivalent circuits*

(*i*) Assuming that there is nothing connected to the termi-
      nals (ie an open-circuit condition) calculate the poten-
      tial differences, V_t and V_n (as functions of R_0 and E_0
      or I_0) between the output terminals of each of the two
      circuits.

(*ii*) Assuming that a wire of zero resistance is used to short-
      circuit each set of terminals, calculate the currents, I_t
      and I_n (as functions of R_0 and E_0 or I_0) that will flow
      through the wire.

      If we put $E = I_0 \times R_0$, what can be said about V_t and V_n,
      and I_t and I_n?

Refer to Fig. 2.3d.

(*i*) with nothing connected to the terminals of each circuit (ie open
      circuit), we have:

$$V_t = E$$

$V_n = I_0 \times R_0$ (since the current flows internally round the
circuit);

(*ii*) With a direct connection of zero resistance between the terminals, (ie short circuit), we have

$$I_l = E/R_o$$

$I_n = I_o$ (since all the current now flows through the external circuit).

If $E = I_o \times R_o$ then it can be seen that

$$V_l = V_n \text{ and } I_l = I_n.$$

The two circuits give exactly the same results. It can be proved that this will still be true if some other resistance is connected to the output. Hence, from outside, both circuits appear to behave in exactly the same way – they are *equivalent*.

We can see now, that it is possible to convert one form of equivalent circuit into another. This is not surprising as they are supposed to be 'equivalent' circuits! It is a matter of convenience as to which is the most useful form to choose for a particular problem.

2.3.4. AC Equivalent Circuits

We can still use these types of circuits if the source of emf, E, and the current, I are AC sources instead of DC sources. See Fig. 2.3e (*i*) and (*ii*).

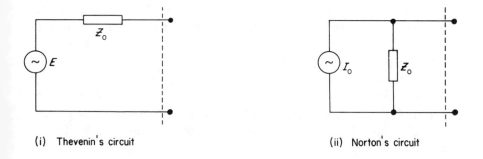

(i)   Thevenin's circuit   (ii) Norton's circuit

Fig. 2.3e.   *Equivalent circuits for AC*

It is important, when discussing AC circuits, to mention the term *impedance*, (Z). This is used to give the ratio between voltage, V, and current, I, in an AC circuit.

$$Z = V/I$$

This looks remarkably like 'resistance' as used in a DC circuit, and indeed it is measured in ohms. However in an AC circuit, there are other factors (capacitance and inductance) which also affect the magnitude of the current and these must be taken into account in addition to pure resistance.

In Fig. 2.3e, Z_o, is called the *output impedance*. Do not worry about using the word 'impedance' instead of 'resistance' – you won't go very far wrong if you assume they are almost the same thing!

2.3.5. Electric Meters

In the previous sections we have been developing circuits that mirror the behaviour of *sources of electricity*. We now want to go to the other end of the problem and look at the way in which we *measure* the electric signal.

One of the simplest forms of electrical measurement involves the use of the *moving-coil meter*, Fig. 2.3f.

The meter consists of a coil of wire wound round an iron former and suspended on a very fine suspension between the poles of a permanent magnet. When the current, I, passes through the coil, the interaction between the magnetic field and the current in the coil causes the coil to turn. The angle through which it turns depends on the magnitude of the current, and the strength of a spring which is opposing the turning motion. The resulting deflection is shown by a pointer moving against a scale calibrated directly in terms of the current, I. When the current is sufficiently high to cause the pointer to travel from one end of the scale to the other, there is a Full-Scale Deflection, FSD.

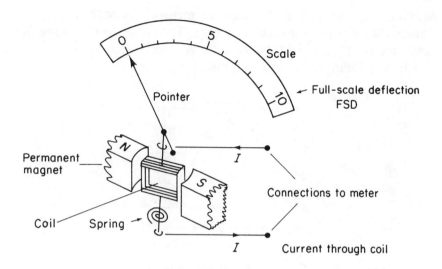

Fig. 2.3f. *Moving-coil meter*

The deflection of the pointer corresponds to the applied *current*. A *perfect* meter would be capable of measuring the current without being a resistance to the flow of that current. In practice, however, all *real* meters have an *internal electric resistance*, R_m. We can represent our *real* meter by an *equivalent circuit*, Fig. 2.3g, consisting of an *ideal* meter, M, and an internal resistance, R_m.

If we want to measure *current* then R_m should be as small as possible.

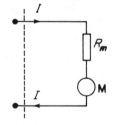

Fig. 2.3g. *Equivalent circuit for meter*

Moving-coil meters are also used to measure *voltage*. How is this done if they normally measure current? The conversion of a moving-coil meter to read voltage is achieved by using an extra load-resistance, R_l, in series with the meter.

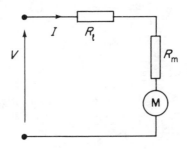

Fig. 2.3h.   *Conversion of meter to measure voltage*

Refer to Fig. 2.3h. If the meter gives a 'full-scale deflection' (FSD) for a current of I_0, then this will occur when the voltage, V, is given by:

$$V = V_0 = I_0 \times (R_l + R_m) = I_0 \times R_i$$

where

$$R_i = R_l + R_m.$$

Thus the meter will show a FSD when a voltage of V_0 V is applied to the 'input terminals' of the circuit shown. The complete circuit is now an instrument for measuring voltages up to V_0. The total resistance, R_i, is called the *input impedance* (or resistance) of the circuit.

If a meter shows a FSD for a current I_0, then, in order to convert it into a voltmeter with a range of V_0 V, an additional resistance must be added so that the *total* input impedance of the combined circuit is given by R_i below.

$$R_i = V_0/I_0$$

∏   A moving-coil meter has an internal resistance of 1000 Ω
     and will give a FSD when a current of 100 μA is flowing.
     Calculate the series resistance, R_l, required so that the meter
     can be used to measure a voltage of 10.0 V with a FSD.

In this question, the meter is required to give a full-scale deflection
(FSD) when a voltage of 10.0 V is applied. We know that the FSD
will occur if a current of 100 μA is flowing. Hence the total resistance
in the circuit, $R_l + R_m$. (see Fig. 2.3h) is

$$= 10.0/(100 \times 10^{-6}) = 1.00 \times 10^5 \ \Omega$$

However, the meter already has its own internal resistance, R_m, of
1000 Ω, hence the extra series resistance required is,

$$R_l = 100 \times 10^3 - 1.00 \times 10^3 = 99.0 \times 10^3 = 99.0 \ k\Omega$$

2.3.6. Electrical Measurement

We now consider the real problem of measurement of an electrical
signal. For the source of our signal we use a Thévenin equivalent
circuit as in Fig. 2.3a, and for our measuring instrument we use a
meter with an input resistance, R_i. If we refer to Fig. 2.3i, the voltage
actually measured will be V, and the current actually measured will
be I.

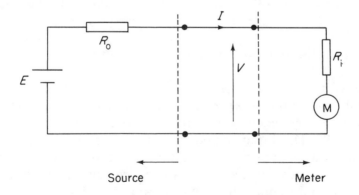

Fig. 2.3i.   *Electrical measurement*

Simple calculation involving Ohm's law gives:

Current $\qquad\qquad I = E/(R_o + R_i)$

Voltage $\qquad\qquad V = I \times R_i = E \times R_i/(R_o + R_i)$

Power dissipated
in the measuring
instrument $\qquad\quad P = V \times I = E^2 \times R_i/(R_o + R_i)^2$

We consider the three different situations as follows.

(*a*) $\quad R_i \gg R_o \qquad$ (put $R_i = \infty$ approx.)

(*b*) $\quad R_i \ll R_o \qquad$ (put $R_i = 0$ approx.)

(*c*) $\quad R_i = R_o$

These situations have the limiting values for I, V and P given in the table below.

Condition	I	V	P	Comment
$R_i \gg R_o$	0	E	0	Measures E
$R_i \ll R_o$	$E/R_o \ (= I_o)$	0	0	Measures I_o
$R_i = R_o$	$E/2R_o \ (= I_o/2)$	$E/2$	$E^2/4R_o$	Maximum Power, P

We can learn three important lessons from this table.

1. If we wish to measure the *true emf*, *E*, of a source then it is essential that the *input impedance of the measuring instrument*, *R_i*, *should be very much greater than the output impedance of the source*, *R_o*. The input impedance of the measuring instrument should be as high as possible so that very little current is drawn from the source. In the ideal case *no current at all* should be drawn from the source by using an input impedance, *R_i*, of infinity. One excellent example of this is in the measurement of the

emf from a pH-electrode. The input impedance of the pH-meter must be considerably greater than the output impedance (*ca.* 1 MΩ) of the pH electrode.

2. If we wish to measure the *maximum current*, I_o available from a source then the *input impedance of the measuring instrument should be considerably less than the output impedance of the source*, R_o. This would be the condition required to measure the current I_o generated in a Norton equivalent circuit. Ideally the input impedance of the measuring circuit should be zero for current measurement. Such a situation occurs when a photovoltaic cell is used to measure light-intensity in a simple spectrophotometer.

3. In our list of three conditions, *maximum power* is transferred from the source to the measuring instrument when $R_i = R_o$, ie when the *output impedance of the source matches the input impedance of the measuring circuit*. It can be proved that this is, in fact, the general condition for maximum power-transfer between the source circuit and the receiving circuit. An example of the use of impedance matching is when a loudspeaker is chosen with an input impedance (say 8 Ω) of the same value as the output impedance (also 8 Ω) of the power audio-amplifier in a 'hi-fi system'.

Π   You are supplied with two voltmeters:

(*i*)   an Avo-meter which has an input impedance of 60 kΩ,

(*ii*) a digital voltmeter with an input impedance of 1 MΩ.

Could either of these be used to make an accurate measurement of the emf from a pH electrode?

Explain the reasons for your answer.

We have seen in the text, and also in Section 2.3.2 that the output impedance of a pH-electrode is of the order of Mega-ohms. If an accurate measurement of the emf is to be made, then the input impedance of the measuring instrument must be considerably greater that 1 MΩ.

Clearly, the Avo-meter with an input impedance of 60 kΩ is quite unsuitable. Its low resistance will have the effect of reducing the measured voltage from the pH electrode to almost zero. To check this you should perform the same calculation as in Section 2.3.2, using 60 kΩ as the external resistance instead of 10 MΩ.

Although it is possible to use certain digital voltmeters connected directly to pH- and ion-selective-electrodes, not all are suitable. If the input impedence is 10 MΩ, this is quite satisfactory for many simple voltage measurements. However, if you compare the situation with that in Section 2.3.2, you can see it is the same calculation, giving the result that there may be an error of 9% in the measured voltage possibly more, depending on the electrode.

Neither meter in (*i*) and (*ii*) can be used directly with a pH-electrode. The required input impedence for meters directly connected to pH- and ion-selective-electrodes should be at least 1×10^{10} Ω! ie 10,000 MΩ.

∏   One form of Infra-red detector uses a *thermocouple* to detect the *heat energy* due to radiation. The thermocouple is made from two dissimilar metals such as platinum and silver. These metals are brought together as very fine wires at a junction which is exposed to the ir radiation. The heat energy in the radiation causes a very slight increase in the temperature of the junction which then produces a small electrical signal.

The power arising from the ir radiation may be very low indeed. Assuming that it is necessary then to transfer as much *power* as possible into the signal processing circuit that follows the thermocouple, estimate the optimum input impedance for the signal processor. The resistance of the thermocouple is 10 Ω.

Choose your answer from one of the following:

(*a*) much greater than 10 Ω

(*b*) approximately equal to 10 Ω

(*c*) zero Ω (a short circuit).

The clue to the answer to this question is in the need to transfer maximum *power*. This corresponds to the last of the three cases considered in Section 2.3.6, where the output impedance of the source should equal the input impedance of the receiving circuit. Here, since the output impedance of the thermocouple will be its own resistance, 10 Ω, then the optimum input impedance of the signal processor should also be 10 Ω – choice (*b*).

SAQ 2.3a

Many instruments have an *analogue output* giving a DC electric signal which can be fed into some other measuring instrument such as a chart recorder. Assume that such an output is giving a DC voltage of 60 mV on open circuit (ie with nothing connected), and that the output impedance of this source is 1 kΩ.

(*i*) Calculate the actual voltage that would be measured if a chart recorder of input impedance 100 kΩ is connected to the output.

(*ii*) Express the difference between the two voltages as a percentage of the open-circuit voltage.

SAQ 2.3b   Assume that the thermocouple used as an
                  ir detector (see Section 2.3.6) has an output
                  impedance of 10 Ω and that, with a given in-
                  tensity of ir radiation, it is producing an emf of
                  7.0 μV.

                  Calculate, for the same intensity of IR radiation,

                  (i)   the maximum voltage signal,

                  (ii) the maximum current,

                  (iii) the maximum power,

                  that would be available if it were possible to
                  change the input impedance of the signal pro-
                  cessor, to which the output thermocouple is con-
                  nected.

2.4. METHODS OF MEASUREMENT

2.4.1. Introduction

There are *three* principle types of measurement procedures:

— Direct Measurement,

— Comparison,

— Substitution.

These categories apply, not only to electrical measurement, but to all forms of measurement. In the following sections we at first differentiate between the three methods by using examples from both electrical measurement and weighing, and then we shall concentrate on one particularly important comparative method –'*null-balance*'.

2.4.2. Direct Measurement

As its name suggests this is the simplest form of measurement. For example, we can measure a voltage by choosing a meter with the correct voltage range, connecting it to the appropriate terminals, and reading the voltage directly from the needle position on the scale. The equivalent method in weighing is to take a spring balance with the correct range, attach the unknown weight, and read the extension against the calibrated scale.

What have these two methods in common? (See Fig. 2.4a).

Both methods rely on the behaviour of some physical system ('transducer' and 'signal processor', see Section 1.2.4) to convert the quantity being measured (the 'input-signal'), into an observable quantity (the 'output signal'). For the voltmeter, the physical process is the rotation of the moving coil when a current flows. The spring balance relies on the extension of the spring caused by the force of gravity on the weight. For each, an initial calibration of the position of the pointer as a function of the magnitude of the 'input signal'

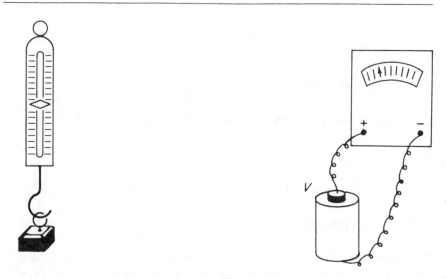

Fig. 2.4a. *Direct measurement*

is required. This is often done only at one position on the scale (typically at full-scale deflection), and the accuracy of the reading at other points then depends on the *linearity* of response of the system. The continuing accuracy of the instrument between calibrations depends on the amount by which the response of the system may change due to ageing and other effects. The accuracy of a *Direct Measurement* quite clearly depends very heavily on what physical system has been chosen as the transducer and the signal processor, how often the system is calibrated, and the general quality of the equipment used.

2.4.3. Comparative Measurement – Null-balance

The *comparative* method of weighing must be very familiar to all! A pair of scales, Fig. 2.4b(i), is used to compare the weights of known and unknown masses. If they are equal there is no deflection of the pointer. If one is greater than the other then there is a deflection to one side or the other depending on the relative mass. It is then merely a matter of having enough known (ie calibrated) weights so that we can be sure of being able *exactly* to balance any unknown weight.

There is no need for calibration. In each measurement the unknown quantity is compared directly with a known quantity.

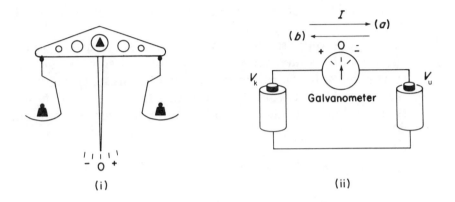

Fig. 2.4b. *Comparative measurement*

A similar situation may occur in electrical measurement. One can produce a known voltage, and then compare this with the unknown voltage, see Fig. 2.4b(ii). The actual comparison is carried out by using a type of ammeter that can detect whether a current is flowing in one direction or the other. This particular type of ammeter is called a *galvanometer*. If $V_u > V_k$, then a current, I, will flow in the direction (b), otherwise in direction (a) for $V_u < V_k$. However, if $V_k = V_u$, then the galvanometer will have no deflection and continue to read zero. When the zero position is obtained (a 'balance'), then provided we know the value of V_k, we also know the value of V_u, which is exactly the same. For obvious reasons, the comparative method described above is called a *Null-Balance* method. The principal reason that the null-balance method is so accurate is that it does not rely on any other physical system (as in direct measurement) to obtain the value of the quantity being measured.

It is true that a simple 'measuring system' (ie the galvanometer or balance pointer) is required, but only to *indicate* when the null-balance is obtained. However, because that system is required only to give a zero reading it does not have to be calibrated, nor does it

even have to give a linear response. The measuring system has to
be calibrated only if readings are taken 'off-balance', as sometimes
happens with the chemical balance. However, these readings represent only the last significant figures in the overall measurement, so it
does not seriously detract from the overall accuracy of the method.

∏   The operation of the basic colorimeter was discussed in Sec-
         tion 1.4.3 in which the percentage transmission of light, T,
         through a solution was measured. If L_o represents the inten-
         sity of light that would pass through the pure solvent, and
         L the intensity of light actually leaving the sample solution,
         then,

             Percentage Transmission, $T = L/L_o \times 100$.

         (We use L here for light intensity instead of the more com-
         mon 'I' to avoid confusion with electric current!)

         In measuring the concentration of solutions, it is more com-
         mon to use a magnitude called *Absorbance*, A, given by

$$A = \log(100/T)$$

         We then read in text books that the concentration, c, of a
         solute in a solution is given by

$$c = A/ba$$

         where b is the path-length of the light through the liquid and
         a is the *absorptivity* of the particular solute. It appears that by
         measuring A by experiment with the colorimeter, knowing b,
         and by looking up a in tables, we can *directly* determine the
         concentration, c, of the solution without using any external
         standards.

         (*i*) Try to think of some reasons why the *direct* method
                given above for determination of concentration may
                give an *inaccurate* result.

(*ii*) Describe how you can use a colorimeter to give a *comparative*, and hence, more accurate, result.

There are very many reasons why the simple colorimeter might give an inaccurate result because of the use of a *direct* method instead of a *comparative* method. The best way of beginning to pick out sources of error is by retracing the calculation and questioning each step. In this case the concentration is given by Beer's law:

$$c = A/ba$$

The first question is: Is this equation correct? For many applications involving dilute solutions it is good enough. However, there are many situations where absorbance A is *not* a linear function of concentration because of some form of physical or chemical activity. Incorrect assumptions made in theory lead to errors in the result.

The next step is to look at the source of each of the values used in our calculation. The absorptivity of the particular solute, a, is taken from tables in a book. An important question here is: Were the conditions used in the measurement of a the *same* as those that you used in your experiment?

Was the same wavelength used?

Was the same range of wavelengths (line-width) used?

Were there any other interacting species in your sample?

There are many different possibilities.

The path length of the light through the sample, b, should be quite easy to measure accurately. However, silly mistakes do occur!

A is the value of absorbance obtained by performing the experiment with the colorimeter. Many factors may make this inaccurate. The response of the detector system may not be *linear* with respect to light intensity, ie a calibration at FSD may not give the correct result for a measurement made at mid-range on the meter. The sensitivity of the colorimeter might change between the time of calibration and of measurement, possibly due to a fluctuation in the brightness of the light source.

It is impossible to give all the reasons that you may have thought of, but what is important is that you realise that errors in this case fall into three main categories:

(*a*) use of an oversimplified theory;

(*b*) the conditions in your experiment may be different from those when the standard was being measured;

(*c*) actual errors in the instrument.

Does your list of errors include items from each of these categories?

The most effective way of improving the accuracy of a simple measurement, such as that made with the colorimeter, is to produce a *calibration curve* for the particular measurement being made. Here it would involve making up a set of standard samples, covering a range of known concentrations (which includes the likely concentration of the unknown sample), and then simply measuring, and plotting on a graph, their apparent absorbances A as a function of the concentrations, c, by using the *same* colorimeter under the *same* conditions as for the sample measurement. The concentration, c, of the sample can then be read from this calibration curve after a measurement of its apparent absorbance, A.

This method avoids problems due to non-linearities in the theory, the conditions of the experiment, and the instrument itself. However, it is frequently tedious to have to draw up a calibration curve for every sample being analysed. Note, however, that this is not the complete solution, because the use of a calibration curve does not compensate for other factors such as the short-term change in the sensitivity of the colorimeter due to a fluctuation in the intensity of the lamp. We shall see later how the use of a double-beam system extends the idea of a 'comparative' method to achieve even greater accuracy!

2.4.4. Measurement by Substitution

We saw in the previous section that the comparative method of measurement is fundamentally more accurate than a corresponding direct method, because of the elimination of the measuring system as a means of interpreting the input-signal being measured. We also saw, however, that a limited form of measuring system was still being used to record the position of null-balance. If we wish to proceed to a yet more fundamentally accurate method of measurement, then we must eliminate any effect of the measuring system.

As an example consider the basic two-pan chemical balance, Fig. 2.4b(i), which is exactly balanced when there is a mass of 200 g on each pan. Now if these weights are removed and a weight of 1 g only is placed on each pan,will there still be perfect balance? Obviously we expect that this will be so. However, between the first and second measurements a total of 398 g has been removed from the system, and this will affect the stresses and strains that are present in the beams, the supports, and the knife edges. It is quite possible that there will be a minute change in the behaviour of the system as a result, and that this may cause a slight displacement from a perfect balance, giving an error in the measurement. In a more accurate balance we would have to make sure that the *total weight* on the system did not change even if we are measuring weights of different values. This can be done by a method of *substitution*.

In Fig. 2.4c, a perfect balance is obtained with 200 g (calibrated weights) in pan *B*. An unknown weight, *M*, is put into pan *A*. In order to come back to balance, it is now necessary to *remove* weights from pan *B*.

Quite clearly, the weight removed from *B* is equal to the unknown weight added to *A*, hence that weight has been measured. However, what is significant in our present discussion is that the *total weight* on the balance *has not changed*. All that has happened is that we have *substituted* an unknown weight for a known weight, and the *conditions of the measurement (the balance) have not changed at all*. Thus measurement by substitution involves *replacing* something of unknown value by something of known value without changing

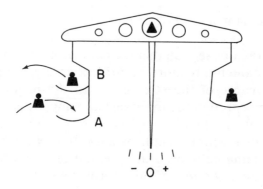

Fig. 2.4c.   *Substitution measurement*

the conditions of measurement. Consider, for example, a resistance of unknown value in a circuit. If we replace that resistance by a resistance of *known value*, R, so that the voltage and current in the circuit remain exactly the same, then the value of the unknown resistance is also equal to R. Measurement by *substitution* is the ultimate test of an unknown value.

It is probably of value to mention here that this also applies to chemical analysis. Suppose that you have performed many determinations on an unknown sample and you are sure that you know exactly what are its constituents, then the most accurate test you can finally perform is to make another sample of exactly the same components and concentration and subject it to all of the determinations (eg measurement of absorbance as above) that you performed on the unknown. If the known and unknown sample give exactly the same results then you know you are correct. It would be very satisfying if your analyses were always sufficiently accurate to be able to do that!

2.4.5.   Potentiometric Measurement

The circuit given in Fig. 2.4b(*ii*) gives the situation where one electrical potential, V_u, is balanced by an equal potential, V_k. This forms the basis for a very important type of electrical null-balance method of measurement – *Potentiometric Measurement*.

In weighing, we have a large set of standard weights which we can balance against an unknown weight. The electrical equivalent would be a large set of accurate and stable standard *potential differences*. However, such a set does not exist, and it is necessary in electricity to generate a range of possible potential differences (voltages) from a *single* standard and calibrated voltage source (standard cell). A circuit which helps us to do this is given in Fig. 2.4d. We assume that the voltage, V_s, is accurately known. Connected to this is a resistance track (as in Fig. 2.1f(*ii*)) with a moving contact, the 'wiper'.

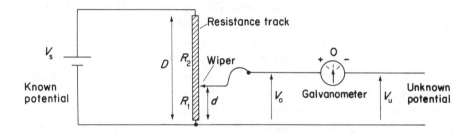

Fig. 2.4d.   *Potentiometric measurement*

We saw in Section 2.1.6 that the voltage V_o is a specific fraction of the voltage, V_s, determined by the relative resistance of the two parts of the track on either side of the point of contact of the wiper. If the resistance track is very carefully made so that the resistance per unit length is constant, then the resistance of each section of the track will be proportional to the length of that part of the track. Therefore,

$$d/D \;=\; R_1/(R_1 \,+\, R_2)$$

Thus

$$V_o \;=\; V_s\, d/D$$

and since both V_s and D are constant, the output voltage, V_o, is then directly proportional to the *distance* of the wiper along the track. Although this is a comparative method, the 'known' voltage is not, itself, a fixed standard value because it must be derived from an electric circuit. The accuracy of the system is limited by the accuracy of the potential divider circuit.

In a typical measurement, the voltage, V_o, could then be compared to an unknown voltage, V_u, by using a galvanometer or some other means to detect when the voltages are exactly equal, see Fig. 2.4d. The distance, d, of the wiper along the track, when the two voltages balance, is directly related to the unknown voltage. If a suitably calibrated scale is attached to the system it is possible to get a direct reading of the unknown voltage. This is the method used in the *potentiometric chart recorder*. The pen which writes on the paper is *directly* connected to a wiper on a resistance track, and a motor drives the wiper (and pen) backwards or forwards until a balance is obtained between V_o and the unknown voltage. An electronic amplifier is used to detect whether the voltages are equal, and this circuit will provide the electronic 'instruction' to drive the motor towards balance. If the output signal from some other system such as a Gas Chromatograph is used as the unknown voltage, then the result is that the pen follows (within limits) the variation of this signal with time, and draws out (in this example) a 'chromatogram' on the paper.

One particular advantage of potentiometric measurement is that when a balance is achieved, *no current is taken from the unknown voltage source, V_u*. This is obviously true because the condition for balance is one of zero current, but why is this good?

We saw in Section 2.3.6 that the optimum condition for measuring voltage is when the input-impedance (resistance) of the measuring circuit is infinite ($R_i = \infty$). This means that the current entering the measuring instrument should ideally be *zero*. This is just the condition that exists for the potentiometric measurement. In many older chemical instruments, it was necessary to obtain a balance condition by hand. The operator was required to look at a meter with a centre-zero reading, and manually adjust a 'pot' control (see Section 2.1.6) until the meter read zero, indicating a balance had been obtained. Most modern instruments have a motorised system to arrive at balance without help from the operator.

∏ In a *double-beam* spectrophotometer, the solution under test and the 'blank' or 'reference' solution (normally pure solvent) each sit in the path of a separate light beam. The double-beam instrument is thus capable of producing electric voltage signals, V_s and V_r, that are proportional to the light intensities (L and L_o respectively) that emerge from the sample and the reference solution.

ie $V_r/V_s = L_o/L$

The voltage, V_r, is applied to the full length of a potentiometer wire (see Fig. 2.4e), and the position of the wiper is adjusted (either manually or automatically) until an electric balance is obtained with V_s.

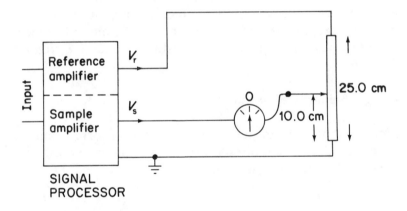

Fig. 2.4e. *Potentiometric measurement*

If the wiper is 25.0 cm long and the position of balance is 10.0 cm as in Fig. 2.4e, calculate the percentage transmission of light through the sample.

Compare the diagrams in Fig. 2.4e and Fig. 2.4d. The known voltage is V_r (from the reference beam) and the unknown voltage is V_s (from the sample beam).

$$D = 25.0 \text{ cm} \quad \text{and} \quad d = 10.0 \text{ cm}$$

Thus

$$V_s = V_r \times 10/25$$

Thus,

$$L/L_0 = V_s/V_r = 0.4$$

Percentage transmission, T, $= L/L_0 \times 100 = 40$.

With the *double-beam* spectrophotometer one can, by comparing a sample and a reference beam, obtain an immediate value for T (and hence A). Compare this with the necessity when using a *single beam* spectrophotometer (like the colorimeter in Section 1.4.3) of manually interchanging sample and reference cuvettes. Using a double-beam system begins to eliminate the problems caused by a fluctuation of the intensity of the light source (see Section 2.4.3), because such a fluctuation will affect the magnitude of V_r and V_s in the *same proportion* and will not then make any difference in their *ratio* (V_r/V_s), thus leaving the measured value of T *unaltered*!

2.4.6.   Bridge Measurement

We can now begin to understand a different form of measurement system by first extending the use of the potentiometric idea developed in the previous section.

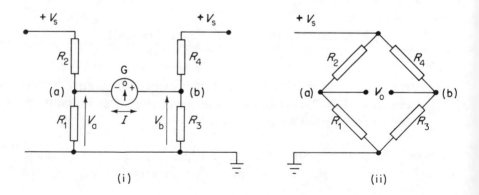

Fig. 2.4f.   *Wheatstone bridge*

Consider Fig. 2.4f(i) where two potential divider chains, are each supplied with a voltage, V_s (which is the same value for each chain). It is possible to calculate the two pd's, V_a and V_b.

$$V_a = V_s \times R_1/(R_1 + R_2)$$

$$V_b = V_s \times R_3/(R_3 + R_4)$$

If there is no current passing through the galvanometer ($I = 0$) then $V_a = V_b$.

A simple piece of algebra then shows that, when this balance occurs, the equations above lead to the following relationship.

$$R_2/R_1 = R_4/R_3$$

Since the voltage, V_s, is applied to both chains of resistors, we can connect them together at the top, giving the more common diagram for the circuit known as the *Wheatstone Bridge* – Fig. 2.4f(ii). The condition necessary to ensure that there is no signal, V_o (or I_o), across the centre of the bridge is given in the equation above relating R_1, R_2, R_3 and R_4. This is the condition for *balance* of the bridge. Quite clearly, the Wheatstone bridge could be used for a straightforward measurement of an unknown resistance, provided that the other three resistors were of known value.

This type of bridge also has a very wide application in many different types of instrument. Consider, for example, the thermal conductivity detector, TCD, discussed in Section 1.3.3. The TCD is used to detect when some volatile organic compound is flowing along with a carrier gas. This is achieved by detecting a change in the *resistance*, R of a heated filament. We are not really interested in the actual resistance, R_o, of the filament, but we are very interested in any *change* in resistance, ΔR. Thus if we set up a bridge with some other balancing resistors we can obtain a null-balance when pure carrier gas only is flowing. When the resistance changes, due to the presence of a sample in the gas, the bridge will come off balance, and there will be an electrical signal appearing at the output of the bridge.

It was proposed in Section 1.3.3 that, in order to avoid having a bogus signal appearing due to a change in carrier flow rate, there should be *two* filaments, both subjected to the same flow-rate of carrier gas, and that only one of these should be in a flow where a sample might be present. These two filaments are represented by their resistances, R_s and R_r, in Fig. 2.4g.

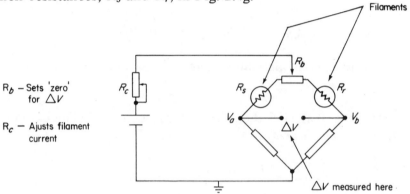

R_b — Sets 'zero' for ΔV

R_c — Ajusts filament current

Fig. 2.4g.   *Two-filament TCD circuit*

When the TCD is set up for operation, the variable resistance, R_c, can be varied to adjust the total current (heating effect) through the filaments so that they reach the desired temperature. It is then necessary to bring the bridge to an initial balance with only gas flowing over both filaments. This balance can be achieved by moving the wiper along R_b until zero output ($\Delta V = 0$) is obtained. We now get a signal, ΔV, from the bridge only when the resistance of the filament in the sample flow, R_s, changes while that of the other or 'reference' filament, R_r, remains constant. This output signal is then a measure of a quantity of sample present in the gas. Note that if the carrier flow-rate changes, then both R_s and R_r are altered by the same amount. The condition for balance of the bridge is still maintained and there is no bogus signal.

We are also interested in measuring resistance in electrochemistry. In a *conductivity* measurement, the conductivity, κ, of the solution in the special *cell* is inversely proportional to the electrical resistance, R, of that cell, and it is possible to measure this by using a 'bridge' method (see Fig. 2.4h).

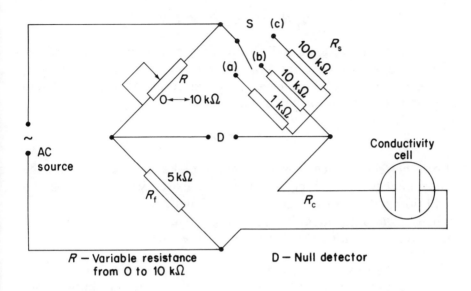

Fig. 2.4h. *Wheatstone bridge for conductivity measurement*

However, a simple DC (direct current) measurement cannot be used because the cell would become polarised by electrolytic action at the electrodes. A bridge measurement can still be used, provided that it is driven by an AC (alternating current) supply. The current is repeatedly changing direction and does not give enough time for polarisation to occur in either direction. With an AC signal, a simple galvanometer cannot be used as the null detector, because it relies on indicating the 'polarity' of a DC signal, and as we have seen in Section 2.2.7, this is not appropriate for an AC signal. Possible alternatives for detection of the amplitude of the AC signal are either the use of earphones or some form of cathode ray display tube. More sophisticated systems also incorporate some form of phase-sensitive detection (see Section 2.2.7) to show in which direction the balance point lies.

In Fig. 2.4h, S is a *Range* switch which allows very different conductivities to be measured with one instrument. The variable resistance, R, is adjusted until the bridge is balanced with zero AC signal at the null detector. If R_c is the resistance of the cell, then the Wheatstone Bridge equation gives,

$$1/R_c = R/(R_s \times R_f)$$

The use of AC instead of DC now means that the balance of the bridge depends on the ratio of *impedances*, Z, (see Section 2.3.4) instead of just the resistances. To be entirely accurate we should write the condition for balance as follows.

$$Z_2/Z_1 = Z_4/Z_3$$

For a *perfect* balance, it is necessary to make an adjustment for the capacitances and inductances that may exist within the bridge, in addition to the main adjustment which balances the pure resistances. This *second* adjustment is sometimes provided on the instrument to improve the quality, or 'sharpness', of the balance.

∏   When using the conductivity bridge as in Fig. 2.4h, the cell is first calibrated by using a standard solution of known conductivity 1.57 S m^{-1} (the units are siemens per metre where 1 S equals 1 ohm^{-1}, alternatively reciprocal ohm per metre, Ω^{-1} m^{-1}). With this standard solution, the range switch, S, is set to position (b) and when the bridge is balanced the value of the resistance, R, is 7.85 kΩ.

    (*i*) Assuming that only the *resistances* in the bridge are significant, calculate the value of resistance, R, required to obtain balance in the bridge when using a solution of conductivity, $\kappa = 0.52$ S m^{-1} in the same cell and also on range (b).

    (*ii*) What will be the three different ranges of conductivity that can be measured by setting the range switch, S, to each of the positions, (a), (b), and (c)?

The condition for balance in the bridge is given by:

$$R_f/R = R_c/R_s$$

However, we are interested in the conductivity, κ of the sample in the cell, and, as we know that resistance is *inversely* proportional to conductivity, it is more useful to write (as in the text),

$$1/R_c \ = \ R/(R_s \times R_f)$$

Thus,

$$\kappa \ = \ k \times 1/R_c \ = \ K \times R/R_s$$

where both k and K (which includes R_f) are constants of proportionality.

During the calibration, we know the values of R, R_s (= 10 kΩ for (b) setting), and κ, which then give

$$1.57 \ = \ K \times 7.85/10.0$$

Thus $K = 2.0$.

(*i*) Thus when $\kappa = 0.52$ S m^{-1}, and with $R_s = 10$ kΩ (range (b)), we have

$$0.52 = 2.0 \times R/10$$

thus $R = 2.60$ kΩ

(*ii*) Variation across each range is given by the possible values of R. The value of R can be altered from 0 to 10 kΩ, see Fig. 2.4h. From the above equation for κ it can be seen that when R is 0 then κ also is 0. This is one end of the range for all settings of the switch S.

For all ranges: $\kappa_{min} = 0$.

When R is at its maximum of 10 kΩ, then the maximum possible value of κ is given by,

$$\kappa_{max} \ = \ K \times R/R_s \ = \ 2.0 \times 10.0/R_s$$

The value of R_s is selected by the Range switch, S, giving each of the three possibilities below.

Range (a)

$R_s = 1.0$ kΩ, thus $\kappa_{\max} = 20$ S m^{-1}.

Range (b)

$R_s = 10$ kΩ, thus $\kappa_{\max} = 2.0$ S m^{-1}.

Range (c)

$R_s = 100$ kΩ, thus $\kappa_{\max} = 0.20$ S m^{-1}.

SAQ 2.4a

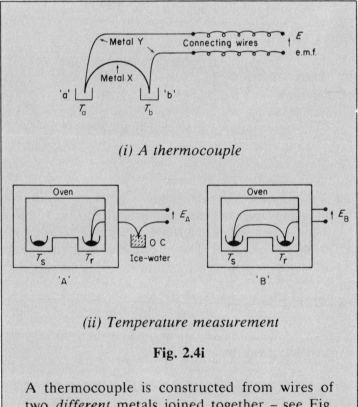

(i) A thermocouple

(ii) Temperature measurement

Fig. 2.4i

A thermocouple is constructed from wires of two *different* metals joined together – see Fig. 2.4i(*i*). If the two junctions, 'a', and 'b', are at different temperatures, T_a and T_b, then an emf

$\longrightarrow$

SAQ 2.4a
(cont.)

(voltage) will be produced in the wire. This emf, E, depends on the *difference* between T_a and T_b, and may often have a *linear* relationship of the form:

$$E = k(T_a - T_b)$$

where k is a constant of the thermocouple. For the purposes of this question use 40 μV °C^{-1} as a typical value for k, ie a temperature difference of 70 °C would produce an emf of 70 × 40 μV = 2.8 mV.

In Differential Thermal Analysis, DTA, a sample is slowly heated in some form of oven or heating block. At the same time, some reference material is also heated in a similar way. Normally the rate at which the temperature of the sample rises will be uniform, except where the sample undergoes some physical or chemical change. This change may be accompanied by the absorption of a large amount of heat or even by the release of heat due to an exothermic reaction. The rate of temperature rise of the reference material would not be affected by such variations. An example is given in the table below of a typical variation in the temperatures of a sample, T_s, and reference material, T_r, when heated in an oven at a controlled rate.

Consider the two situations depicted in Fig. 2.4i(ii). In each situation one junction of a thermocouple is placed in the sample, but the other junction is placed as in the two situations below.
$\longrightarrow$

SAQ 2.4a
(cont.)

'Situation A', the other junction is placed in an ice-water mixture to maintain a constant temperature of 0 °C: thermal analysis.

'Situation B', the other junction is placed in the reference material at the varying temperature, T_r: differential thermal analysis.

(*i*)   Plot on the two graphs provided [Fig. 2.4j(*i*) and (*ii*)] the variation of the emf's produced in the two situations over the complete temperature range. Use $k = 40$ $\mu V\ °C^{-1}$ as above.

Note the *different abscissae* for the two graphs.

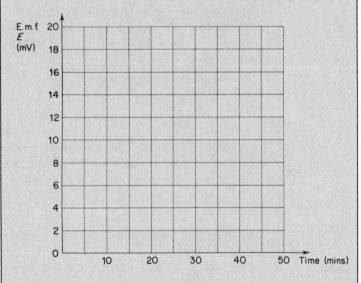

(i) Situation 'A'

Fig. 2.4j

$\longrightarrow$

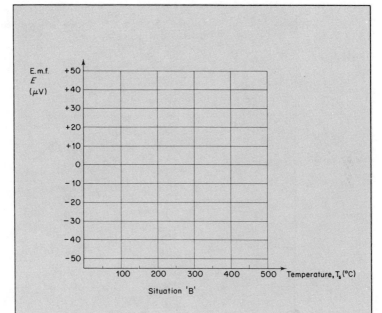

(ii) Situation 'B'

Fig. 2.4j (cont.)

(*ii*) Calculate, from the graphs, the average rate, in °C min^{-1} at which the temperature of the oven is rising.

(*iii*) Identify points where a fluctuation occurs in the heating curve of the oven.

(*iv*) Determine at what temperature some form of physical or chemical change occurs in the sample.

(*v*) Comment on the advantages of placing one junction of the thermocouple in the reference material. $\longrightarrow$

Table of results.

Time (min)	T_r (°C)	T_s (°C)
0	50.0	50.0
5	100.0	100.0
10	150.0	150.0
11	160.0	160.0
12	170.0	169.5
13	180.0	179.0
14	195.0	194.5
15	210.0	210.0
17	240.0	240.0
19	265.0	265.0
21	280.0	280.0
23	290.0	290.0
25	300.0	300.0
27	320.0	320.0
29	340.0	340.0
31	355.0	355.25
32	360.0	360.5
33	365.0	365.25
34	370.0	370.0
36	385.0	385.0
38	410.0	410.0
40	440.0	440.0
42	470.0	470.0
44	490.0	490.0
45	500.0	500.0

SAQ 2.4a

SAQ 2.4b   Assume that a circuit as in Fig. 2.4g, with semi-conductor filaments, is used for a TCD system, and that, for a certain flow-rate of helium carrier gas, the two voltages, V_a and V_b, are equal in the absence of any sample in the gas. Indicate below the way in which these two voltages may change if the flow-rate of helium *decreases*. (Note that this question is concerned only with a *change* of *flow-rate* and not with the presence of any sample in either of the two gas flows.)

If you are in any doubt about the behaviour of the TCD detector, refer back to Section 1.3.3. Note that the positive potential is being applied to the *top* of the bridge in the diagram.

Possible responses.

(*i*) V_a increases and V_b decreases,

(*ii*) V_a increases and V_b increases,

(*iii*) V_a decreases and V_b decreases,

(*iv*) V_a decreases and V_b increases.

SAQ 2.4b

Summary

The first section reviews the basic principles of electricity by working through problems of particular relevance to instrumentation. We then discuss the possible roles of electricity as a provider of motive power and/or a conveyer of information.

The concept of the 'equivalent circuit' is developed as a foundation to a discussion of the problems of measurement in electrical systems. Emphasis is placed on understanding the importance of the input- and output-impedances of systems whenever they are connected together (eg connecting a pH-electrode to the pH-meter).

Finally, the problems of electrical measurement are then expanded into a consideration of measurement techniques in general.

Objectives

It is expected that, on completion of this Part, the student will be able to:

- use the concepts of electrical potential, current, and resistance, for both DC and AC;

- appreciate the use of electricity as a carrier of power and/or information;

- perform calculations on signal waves involving amplitude, frequency, and phase difference;

- use the concept of output- and input-impedance in calculating the effect of connections between instruments;

- apply the concept of differential and comparative measurements.

3. Signals and Signal Processing Elements

Overview

The nature of signals and of 'noise' in both the frequency and the time domain is introduced, leading to a discussion of the principal types of analogue signal-processing units. The treatment is intended to convey the concepts of the processes rather than entering into detailed electronics. The factors affecting the signal-to-noise ratio are an essential part of this treatment.

3.1. THE NATURE OF SIGNALS

3.1.1. Introduction

We have already seen some of the properties of *electric* waves in Section 2.1.9. We introduced there the terms amplitude, frequency, period, and phase of sinusoidal waves. For the chemist, another important example of wave motion is, of course, *electromagnetic waves*, eg light waves, X-rays, infra-red (ir), and ultra-violet (uv) radiation. The electromagnetic (em) spectrum will be dealt with elsewhere, but it is useful here just to remind ourselves of the basic properties of any wave motion.

3.1.2.   Waves

A wave, A, is drawn in Fig. 3.1a as a graph of *displacement* against *distance*. The 'displacement' depends on the type of wave: it may be the magnitude of an electric field (as for em waves), the pressure of air (as for sound), or voltage (as in an electric circuit). Notice that we are using the word 'displacement' as a general concept to describe the instantaneous value of some changing quantity.

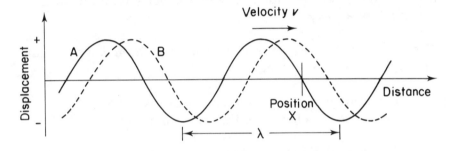

Fig. 3.1a.   *Travelling wave*

You should note however, that the abscissa in Fig. 3.1a is now *distance* and not *time* as in Section 2.1.9. The distance between points of the *same phase* in a given wave (eg between one trough and the next) is called the *wavelength*, λ. Compare this with the *period*, *T*, when we used time as the abscissa (in Section 2.1.9).

We assume that A is a wave travelling with uniform *velocity*, *v*. To illustrate the point we have drawn a second wave B, which represents the position that A will reach after a short time. The time it will take for the wave to travel a distance equal to its own wavelength is, *T*, where

$$T = \lambda/v \qquad\qquad (3.1)$$

During this time one complete wave will pass a stationary observer (eg at position X), and in this time he would observe that the displacement in the wave goes from zero, through a crest and a trough and back to zero again – ie one complete cycle. Thus *T* is also the *period* of the wave.

The frequency, f, of the wave is now given by

$$f = 1/T = v/\lambda \qquad (3.2)$$

or

$$v = f\lambda \qquad (3.3)$$

This is an important equation relating the velocity, frequency, and wavelength of a wave.

All electromagnetic waves travelling in a vacuum have the same velocity, c, viz 3.00 × 10⁸ m s⁻¹.

Π   The frequency of a particular ir wave is 3.0×10^{13} Hz. Calculate the wavelength of this wave in a vacuum.

Will uv radiation have a longer or shorter wavelength? Guess the answer if you do not already know it.

We must use the relationship

$$v = f\lambda$$

We know that *all* electromagnetic waves travel with the same velocity in a vacuum.

$$\therefore \qquad v = c = 3.00 \times 10^8 \text{ m s}^{-1}$$

This gives us

$$\lambda = v/f = 3.00 \times 10^8 / 3.0 \times 10^{13} = 1.0 \times 10^{-5} \text{ m} = 10\mu\text{m}.$$

Provided you were able to find the velocity, you should have had no difficulty with the first part of this problem.

The wavelengths of uv radiation are much smaller than 10 μm. Uv radiation has wavelength < *ca.* 350 nanometres (= 0.350 μm).

The wavelengths of visible radiation extend from about 350 nm to about 700 nm.

You may already know that the wavelength of ir radiation can be used to measure the dimensions of a cell used in ir spectroscopy. Although this involves the phenomenon of interference, which is outside the scope of this unit, we can say that it works only if the wavelength of the radiation is not too small compared to the size of the cell. This technique of interference measurement cannot be used for cells used in visible or uv spectroscopy because the wavelengths are far too small compared to the size of the cell.

3.1.3. Frequency Domain and Time Domain

We saw in Section 2.2.7 that the characteristics of a sinusoidal voltage wave could be defined by using the frequency, phase, and amplitude (maximum displacement) of the voltage signal. If we forget 'phase' for the moment then we can plot the characteristics of a given wave in a graph by using the amplitude (voltage) and frequency as the axes of our graph – see Fig. 3.1b(i).

The single vertical line represents a pure sine wave of frequency, f_0, and amplitude, V_0. The traditional 'picture' of the same wave expresses displacement as a function of time – Fig. 3.1b(ii). We can see that the two parts (i) and (ii) of Fig. 3.1b contain the same *information*, ie the *amplitude*, V_0, and the *frequency*, f_0, of the wave.

Certainly we are more familiar with the representation in (ii), but in fact it does not tell us anything more (apart from phase) than the representation in (i). The 'picture' in (ii) is a representation of the wave in the *Time Domain* (ie using time as an axis) and that in (i) is in the *Frequency Domain* (ie using frequency as an axis). Sometimes we shall find it more useful to use representations in the frequency domain than in the time domain. We shall meet some examples of this in the following sections.

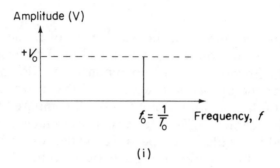

(i)

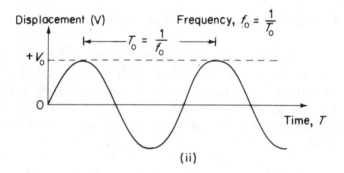

(ii)

Fig. 3.1b.   *Pure sine wave*

3.1.4.   Fourier Transforms

Consider what happens if we start with a sine wave of frequency, f_0, and add another wave of frequency, $3f_0$. At every point in time the displacement from each wave can be simply added together. The resultant combined wave is given in Fig. 3.1c(i), in which we can see that the rising part of the wave has been made steeper, and that the peak of the wave has been made flatter. If now a further frequency, $5f_0$, is added, this continues to flatten the peak and make the rise and fall sharper – see Fig. 3.1c(ii).

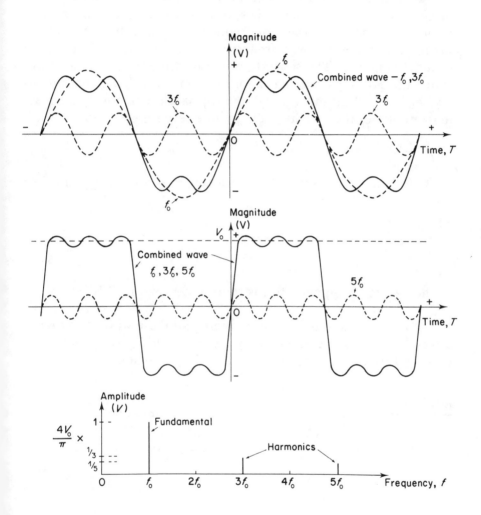

Fig. 3.1c. *Addition of selected frequencies*

These particular diagrams are very complex when presented in the time domain, but you can see in Fig. 3.1c(*iii*) that the addition of extra waves of different frequencies can be achieved in the frequency domain by simply adding a new vertical bar at each appropriate frequency. The basic wave of frequency, f_0, is called the *fundamental* and the waves with frequencies equal to multiples of f_0 are called *harmonics*. The separate waves of different frequencies are called the *Fourier Components* of the transform. If we continue to add higher odd harmonics with appropriate progressively decreasing amplitudes then we have the infinite series given below.

$$V(t) =$$
$$4\,V_0/\pi\,(\sin \omega t + \frac{1}{3}\sin 3\omega t + \frac{1}{5}\sin 5\omega t + \frac{1}{7}\sin 7\omega t + \ldots) \quad (3.4)$$

where

$$\omega = 2\pi f_0$$

In the frequency domain, this is represented by Fig. 3.1d(*i*).

In the time domain, the result of adding all these waves together is a wave with an increasingly steep rise and fall together with a flat crest and trough, ie a *square wave*, see Fig. 3.1d(*ii*).

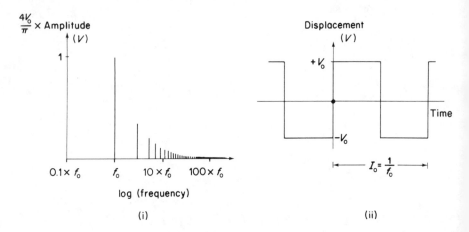

Fig. 3.1d.   *Fourier components of a square wave*

Fig. 3.1d(i) is said to be the *Fourier Transform* of Fig. 3.1d(ii). Similarly, Fig. 3.1d(ii) is the *Fourier Transform* of Fig. 3.1d(i). These Fourier Transforms are just different ways of presenting the same *information* (except phase). Neither is more correct than the other: we are, however, more familiar with the presentation in the time domain. We shall see below how useful the frequency domain presentation can be in understanding the behaviour of signals. Although we have obtained the Fourier Transform (FT) of the relatively simple example of a square wave, it is in fact possible to obtain the FT in the frequency domain for *any complicated signal* in the time domain.

∏   We have seen how a square wave (Fig. 3.1d) can be considered as the sum of many sinusoidal components.

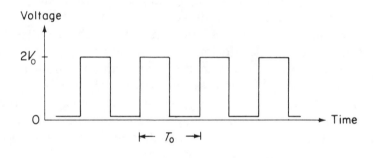

Fig. 3.1e

What additional component must be added to obtain the waveform in Fig. 3.1e ?

The only fundamental difference between the square wave in Fig. 3.1d, and that in Fig. 3.1e is that in the latter a constant voltage V_0 has been added to *every* point in the wave. Thus the extra *component* that must be added is at zero frequency, $f = 0$, ie a DC component. Although we have previously implied a fundamental difference between AC and DC signals, when we come to Fourier Analysis, a DC signal is merely an additional component at zero frequency.

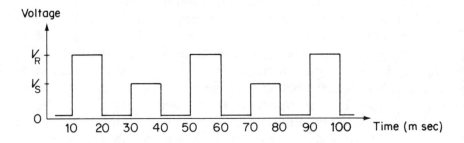

Fig. 3.1f

Π   Consider now the waveform in Fig. 3.1f. We already know
     from the previous exercise that there is a component at zero
     frequency. In this question you are asked first to find the *fre-
     quencies* of the *two* (AC) components with the *lowest* fre-
     quencies. You should find one frequency very easily, but if
     you have problems with the next frequency note that the
     square wave pulses are of alternating height. Having found
     the two low-frequency components, can you explain how the
     amplitudes of these two components might be related to the
     amplitudes, V_R and V_S, in the diagram? Do not spend too
     long on this part if you find it difficult.

If we ignore at first the difference in height of different pulses, the
period of the square wave becomes $T_2 = 20$ ms. This gives us a
frequency $f_2 = 1/T_2 = 50$ Hz – see Fig. 3.1g.

The next step is more difficult. This is where we try to take account
of the difference sizes of the pulses. The variation of pulse size is
sinusoidal with a period, T_1.

The period T_1 can be calculated to be 40 ms; this then gives a frequency, $f_1 = 1/T_1 = 25$ Hz.

The amplitude of the component f_2 must depend on the *average*
amplitude of the pulses, ie $(V_R + V_S)/2$. The amplitude of component f_1, depends on the difference between V_R and V_S ie $(V_R - V_S)$.

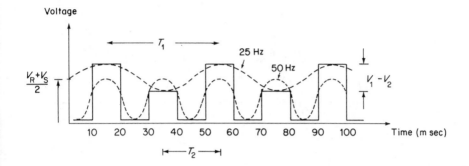

Fig. 3.1g

You should see that this is consistent with our analysis of the square pulse, because the f_1 component completely vanishes for a square wave, ie for $V_R = V_S$.

In terms of *information* the f_2 component gives the *average* amplitude of the signal, and the f_1 component gives the *difference* between the amplitudes of the pulses.

A complete Fourier analysis would also give a range of higher frequency components, but these contain information only about the exact *shape* of the pulses and do not tell us anything new about their amplitude.

Before proceeding it is useful to look at the abscissa of the FT in Fig. 3.1d(i). You can see that we have used a *log f* scale. A logarithmic scale is often used because the *variation* in behaviour of a circuit for two different frequencies, f_a and f_b, usually depends on the *ratio* f_a/f_b rather than the simple *linear difference*, $f_a - f_b$. The logarithmic scale expresses the *ratio* f_a/f_b as the *linear function*, $\log f_a - \log f_b = \log f_a/f_b$, which is then plotted as the abscissa.

An *octave* is a distance on the $\log f$ scale which describes a ratio of *two* between frequencies (as in a musical scale), and a *decibel* describes a ratio of *ten*. In Fig. 3.1h, we have drawn a $\log f$ scale with examples of octave and decibel intervals. Note that on a log scale the *zero* of frequency is infinitely far to the left, $\log 0 = -\infty$.

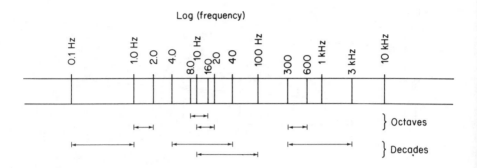

Fig. 3.1h. *Logarithmic frequency scale*

∏   Calculate the frequencies that are:

(*i*)   a decade higher than 30 Hz,

(*ii*)   a decade lower than 3 Hz.

(*i*)   A decade means a *ratio* of 10. Hence a decade higher than 30 Hz will be 30 × 10 = 300 Hz. If you got 40 Hz as an answer, you made a simple addition of frequencies instead of using the multiplication factor.

(*ii*)   Similarly a decade lower than 3 Hz will be 3/10 = 0.3 Hz. It may seem odd to have a frequency of less than 1 Hz. Instead we could give the period (3.33 s in this case), but there is no fundamental reason why it should not be expressed as a frequency of less than 1 Hz.

We can use the idea of Fourier Transforms to help us to understand why it is that when on a flute and a clarinet we play the same note (for example 'A', with a frequency of 444 Hz) the two instruments actually sound different. We are still able to distinguish which one is playing.

The reason is that, although the *fundamental*, f_0, is the same (444 Hz), each instrument has a different mixture of *harmonics* – the *shapes* of the FTs are different – see Fig. 3.1i. It is fortunate for our enjoyment of music that our ears are capable of analysing the FT of the sound signal and our brain is then able to distinguish between instruments!

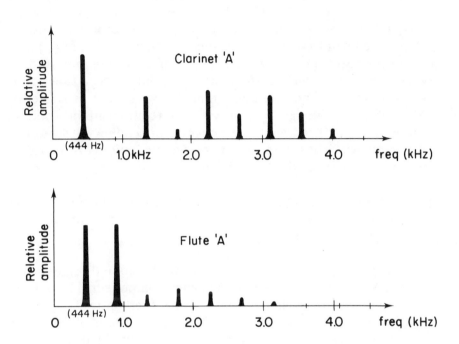

Fig. 3.1i. *Fourier frequency components of a note from clarinet and from flute*

The use of Fourier Transforms appears in several areas of modern chemical instrumentation, eg nmr and ir spectroscopy. It is here more efficient to analyse the FT of a signal than the signal itself. These advanced applications will be mentioned again in Section 5.1.

3.1.5. Spectra

There are a number of very fundamental concepts in science and mathematics which keep re-appearing in different forms. The idea of a Fourier Transform is just one of these. We saw how the FT of the sound of a musical instrument had a shape characteristic of that particular instrument. A similar situation exists with colours in light. You are aware, for example, that there are many different shades of colour which could still be called 'red'. You probably also know that if we look at the *spectrum* of the light then the different shades of red will have a different mixture of colours from different parts of the spectrum. The spectrum is a graph of the amplitude of the various components in the light signal plotted as a function of their wavelength. Since we know that the wavelength is related to the frequency of the light by a very simple relationship, it would be a simple matter to redraw the graph with a frequency instead of a wavelength axis. Thus the familiar spectrum of colours is in effect a *Fourier Transform* of the light signal.

∏   This exercise is concerned with the way in which the human eye can analyse the Fourier Transform (Spectrum) of a light signal.

The derivation of the spectrum can be done directly in *two* main ways: (*a*) different frequency components in the spectrum can be isolated from one another optically by using some filter mechanism, and then the amplitude of each component can be measured one after the other by using a single detector; (*b*) several detectors can be used which are each sensitive to different components of the spectrum.

In the majority of spectrophotometers we use the former method. A single detector measures the amplitude of a particular wavelength component selected by a monochromator, and, by driving the monochromator through a range of wavelengths, the whole spectrum is analysed.

The human eye, however, uses the latter method with three colour sensors, sensitive respectively to *red*, *green*, and *blue* light.

Why is it better for us that the eye uses the latter method?

When looking at a scene we need to be able to obtain a *simultaneous* response at different wavelengths. If we were watching a fast-moving scene, the whole picture would become confused if there was a time delay between receiving each of the three different colours. The eye uses several sensors for exactly the same reason that some hplc uv-visible detectors use more than one sensor. In the hplc uv-visible detector we may wish simultaneously to monitor the absorbance of more than one wavelength of light by the sample as it passes out of the hplc column.

Simultaneous detection of many different wavelengths is now becoming possible with development of *diode array detectors*. These are many individual sensors formed side by side on an electronic 'chip'. The outputs from the sensors can be analysed almost simultaneously in a computer system.

In addition to the visible spectrum referred to above, spectra occur in very many forms. There is always a plot of the amplitude of some signal as a function of another variable parameter. We can stretch the exact meaning of spectrum (which is concerned with light) to cover similar situations in other branches of chemistry, as exemplified below.

Type of 'spectrum'	Variable parameter	Units
Visible spectrum	Wavelength	Nanometres
Ir Spectrum	Wave number	1/Centimetres
Gamma-ray spectrum	Energy	Electron-volts
Gas chromatogram	Time	Seconds
Thermogram (thermal analysis)	Temperature	Kelvins

We are always interested in the information conveyed by the amplitude of the various components at different values of the variable parameter. When information is presented in the form of a spectrum (in its widest sense), the interpretation of that information is subject to similar theoretical and instrumental factors. Terms such as 'line-width', 'peak height', 'baseline drift', 'response time', 'noise', may appear in connection with all types of spectral output from an instrument. You should always try to relate what you learn about the interpretation of one sort of spectrum (eg ir) to that of another spectrum (eg gc).

3.1.6.   Fourier Transform of an Output Signal

A signal of the type shown in Fig. 3.1j(i) can occur in the output of almost any type of instrument. We shall assume, as is often true, that the signal has been drawn out in a chart recorder, and that the magnitude of the signal is given by a voltage as shown.

The signal may represent a single spectral line in the absorption spectrum from a spectrophotometer; the time taken to draw out the line then depends of the speed at which the spectrophotometer is sweeeping through the wavelengths. Alternatively if this is the output from a Gas Chromatograph then the time taken depends on the sample and the settings (oven temperatures) of the instrument.

The width of the signal in time, Δt, depends on the speed at which the instrument draws out the signal. This is usually under the control of the operator, eg by varying the scan speed in a spectrophotometer.

A *single* spectral line is not a repetitive (periodic) signal like the square wave discussed earlier. For such non-periodic waves the FT is no longer an accumulation of *separate* frequency components (as in Fig. 3.1d), but a continuous curve representing contributions from all possible frequency components within a certain range. The Fourier Transform of the spectral line is given in Fig. 3.1j(ii). The lowest frequency component in the transform is a DC component, ($f = 0$). It is however, impossible to identify an exact upper frequency limit to the range of components, because the contributions from these components gradually tail off at high frequency and there is no sharp cut off. Nevertheless, we can use the concept of 'width at half-height' to give an idea of the spread of the signal in time (line width), Δt, and the spread of the FT in frequency (band width), Δf. For the spectral line shown, which we have assumed has a Gaussian shape, we can relate the spread in the frequency spectrum to the spread in the time spectrum by the approximate relationship below.

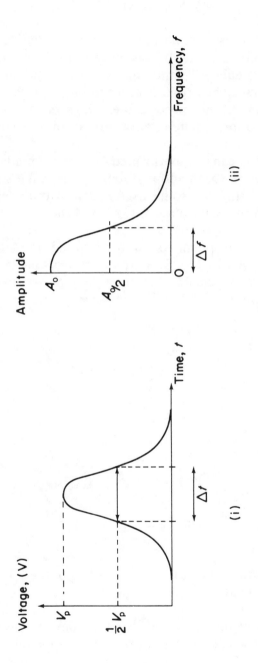

Fig. 3.1j. *Output 'line' signal*

$$\Delta f \simeq 0.5/\Delta t \tag{3.5}$$

Thus if our signal line is very narrow (Δt small) then the spread of frequency components in the FT becomes large, ie a narrow line width requires high frequency components to carry the same information, and *vice versa*. The actual *shape and band width* of the FT depends on the *shape* of the spectral line. However the expression above gives a useful approximate relationship for most applications.

Calculations of FT's can be performed with full mathematical rigour. However the approach adopted here is somewhat simplified and will, we hope, bring out some useful ideas,without offending those readers who already have a good grasp of the subject.

∏   An absorption spectrum containing a broad line and a narrow line is given in Fig. 3.1k. It was drawn out by a spectrophotometer which was sweeping through wavelengths at a rate of 10 nm s^{-1}.

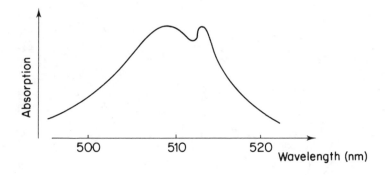

Fig. 3.1k

Make a rough estimate of the bandwidths of the FTs of each line in that signal as it was being drawn.

If the sweep of the spectrophotometer was increased to 20 nm s^{-1}, what would be the effect on the band width?

The first step is to convert the abscissa into seconds by using the sweep rate of 10 nm s^{-1}. Sweeping from 500 nm to 520 nm will take 2.0 s.

The next step is to look at the *structure* of the signal. We can see that it is made up of two 'lines'; a narrow line b, superimposed on the shoulder of a broad line, a. Separating these two and expressing the abscissa in seconds we get the lines in Fig. 3.11.

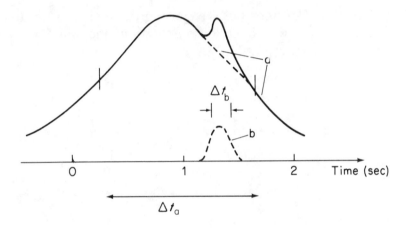

Fig. 3.11

The widths at half-height for the two lines, Δt_a and Δt_b, are about 1.1 s and 0.15 s respectively. By using the relationship in the text ($\Delta f \simeq 0.5/\Delta t$) we obtain band widths as follows.

$$\Delta f_a \simeq 0.50/1.1 \simeq 0.45 \text{ Hz}$$

$$\Delta f_b \simeq 0.50/0.15 \simeq 3.3 \text{ Hz}$$

If the sweep-speed were increased to 20.0 nm s^{-1}, the time to draw out the lines would be reduced by a factor of two. A reduction of the period corresponds to an increase of frequency. Hence we find that:

$$\Delta f_a' \simeq 0.9 \text{ Hz}$$

$$\Delta f_b' \simeq 6.6 \text{ Hz.}$$

If you are not happy about this last step, then you should repeat the initial calculation using 20.0 nm s^{-1} instead of 10 nm s^{-1}, and see how the factor of two follows throughout the calculation.

SAQ 3.1a

Consider the waveform given in Fig. 3.1m and answer the following questions about its Fourier Transform.

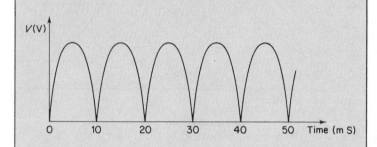

Fig. 3.1m

(*i*) Is there a DC (zero frequency) component in the transform?

(*ii*) What is the lowest (AC) frequency in the transform?

SAQ 3.1b

The waveform in Fig. 3.1n is an Electro-Cardiogram recording of a human heartbeat. You can see that there are five main signals contained in the single beat.

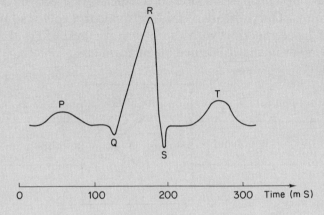

Fig. 3.1n

Estimate the bandwidths of the frequencies contained in each of the five signals.

What do you think would happen, in general terms, to the shape of the heart-beat signal if it were processed by an instrument which could cope only with a bandwidth of 30 Hz?

3.2.   AMPLIFICATION AND ATTENUATION

3.2.1.   Introduction

We do not propose here to introduce any electronics or electronic circuitry. The purpose of this section is to explain some of the terms used in connection with the electronic amplifiers that the chemist may meet in the literature on instruments.

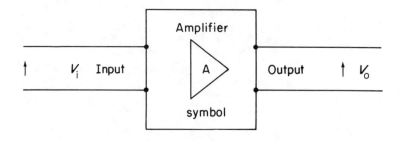

Fig. 3.2a.   *Voltage amplifier*

We shall discuss the amplifier shown in Fig. 3.2a which has terminals for an input voltage and terminals for an output voltage. An electric signal, of voltage V_i enters the amplifier at the input terminals and an electric signal, of voltage V_o, is produced at the output terminals.

3.2.2.   Gain

The voltage *gain* of the amplifier, G, is given by, Eq. 3.6.

$$G \;=\; V_o/V_i \tag{3.6}$$

If for example, the input signal has a magnitude of 3.0 mV and the output signal a magnitude of 81 mV, then the gain of the circuit is $81/3 = 27$. It is quite common for an amplifier also to *invert* the signal, ie a positive input will give a negative output and *vice versa*. For example the 3.0 mV input signal may result in an output signal of -81 mV, giving a gain of $-81/3 = -27$. Even though the gain is negative, the circuit would still be called an amplifier because the absolute magnitude of the output is still greater than that of the input. It would, in fact, be called an *Inverting Amplifier*.

If a circuit has a gain whose *absolute* magnitude is <1, ie

$$|G| < 1$$

then the effect is a reduction in the size of the signal. This is called *attenuation*.

The expression for the gain given above (Eq. 3.6) is a simple ratio of the input and the output voltage. It is also fairly common practice to express the gain as a logarithmic function of this ratio, as given below.

$$G(dB) = 10 \log (V_o/V_i)^2 \qquad (3.7)$$

$G(dB)$ is said to be the Gain in *decibels*. The reasons for choosing an expression of this form do not concern us here, except to note that since the voltage is squared we are really looking at a ratio of input and output *power*.

Important points to note are:

(*a*) an amplifier (inverting or non-inverting) will have $G(dB) > 0$, ie positive,

(*b*) an attenuator will have $G(dB) < 0$, ie negative.

For other comparisons see the exercise below.

NB. We shall normally use G to represent the gain expressed as a ratio, and $G(dB)$ for the gain in decibels.

∏   The table below lists the input and the output voltage of several circuits, together with the gains expressed as ratios and in decibels. Several entries have been missed out.

Complete the table, with calculation where appropriate, by filling in the necessary entries.

	V_i	V_o	Gain, G	$G(dB)$	Circuit type
(*i*)	2.00	200	100	40	Non-inverting amplifier
(*ii*)	200	2.00	0.01	−40	Attenuator
(*iii*)	2.00	−200	...	...	...
(*iv*)	...	100	−20	...	...
(*v*)	...	40.0	...	6	Inverting amplifier

The completed table should look like this.

	V_i	V_o	Gain, G	$G(dB)$	Circuit type
(*i*)	2.00	200	100	40	Non-inverting amplifier
(*ii*)	200	2.00	0.01	−40	Attenuator
(*iii*)	2.00	−200	−100	40	Inverting amplifier
(*iv*)	−5.00	100	−20	26	Inverting amplifier
(*v*)	−20.0	40.0	−2.0	6	Inverting amplifier

You should have found rows (*iii*) and (*iv*) relatively easy because it is possible to calculate G and V_i respectively by using the definition of gain given in the text. The gain (*dB*) is calculated directly from the gain expressed as a ratio.

You may have found problems with row (*v*) in which you must calculate the gain as a ratio given $G(dB)$. If you are not familiar with logarithms you may have made a simple mathematical slip. Check this again if you were wrong.

The type of circuit can be identified from the value of the gain. Note however, that $G(dB)$ does not distinguish between an inverting and non-inverting amplifier. This is because in deriving $G(dB)$ the gain as a ratio is squared thereby eliminating the minus sign.

3.2.3. Band-width

In the previous two sections we have used the term 'gain' as though a particular amplifier or attenuator had a single value for its gain.

However we have seen in Section 3.1.4 that a typical signal may well contain many different Fourier components, and we must ask ourselves whether or not the same value of the gain will apply to all the different components. The answer to that question for a particular amplifier can be expressed by using a graph of Gain, G, *versus* frequency, f as in Fig. 3.2b.(i)–(iii).

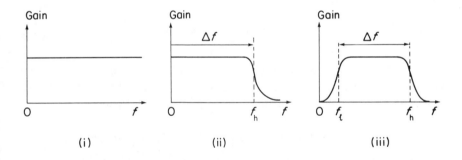

Fig. 3.2b. *Frequency response*

In Fig. 3.2b(i) the gain is seen to have the same value at all frequencies. In practice that ideal situation is impossible for a great variety of technical reasons. An amplifier that will function for DC and at very low frequencies will not be able to amplify a signal at frequencies as high as, for example, 2 GHz (2×10^9 Hz). Thus we expect to see a gain function as in Fig. 3.2b(ii), where there is a high-frequency cut-off point, f_h, at which the gain falls away. Many amplifiers will also have a *low-frequency cut-off* point, f_l, below which the gain falls away, as in Fig. 3.2b(iii).

The gain does not suddenly drop from its maximum value to zero, but gradually tails off either to high or to low frequencies. How then do we define the cut-off points? The definition usually accepted is the point at which the gain as a ratio has dropped to $1/\sqrt{2}$ of its initial value. This is equivalent to a reduction in gain of 3 dB.

These points are then referred to as the '3 dB down-points', or 'half-power points'. The *band-width* Δf, of the amplifier is then defined as the range of frequencies between the low-frequency cut-off and the high-frequency cut-off point, ie

$$\Delta f = f_h - f_l \tag{3.7}$$

A *DC amplifier* must obviously amplify DC voltages. The variation in gain for a DC amplifier is shown in Fig. 3.2b(*ii*) in which there is no low-frequency cut-off and the graph extends right down to DC ($f = 0$). The amplifier will also amplify AC signals up to the high-frequency cut-off point, so that the band-width of the DC amplifier is given by Eq. 3.8.

$$\Delta f = f_h - 0 = f_h \tag{3.8}$$

There is sometimes some confusion about the difference between AC and DC amplifiers. In fact they will both amplify some AC signals, but only the DC amplifier will amplify a DC signal.

The amplifier which has a gain as in Fig. 3.2b(*iii*) is unable to amplify very low frequencies below f_l, and it is therefore unable to amplify a DC signal (which has an effective frequency of zero). Being limited to AC signals only, this amplifier would be called an *AC amplifier*.

We can assign a band-width to systems other than electronic amplifiers. The adult human ear can respond to sounds with frequencies in the approximate range of 20 Hz to 15 kHz. The human eye can respond to radiation with electromagnetic frequencies from about 4.3×10^{14} Hz to 8.6×10^{14} Hz.

∏   An amplifier has a gain/frequency response as in Fig. 3.2c.

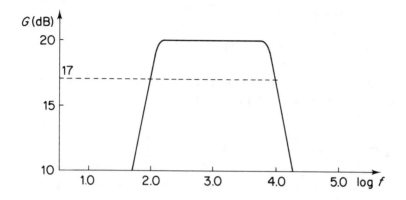

Fig. 3.2c

Derive from the graph

(*i*)   the high frequency cut-off point, f_h,

(*ii*)   the low frequency cut-off point, f_l,

(*iii*) the band-width, Δf.

We should use the definition given in the text for the high-frequency and the low-frequency cut-off point, ie the points at which the gain drops by 3 dB.

The maximum gain is 20 dB thus the cut-off point will come when this drops to $20 - 3 = 17$ dB.

This gives

$$\log f_h = 4.0$$
$$f_h = 10000 \text{ Hz}$$

$$\log f_l = 2.0$$
$$f_l = 100 \text{ Hz}$$

The band-width is given by

$$\Delta f = f_h - f_l = 9900 \text{ Hz}$$

If you have the wrong answer you may have been confused by the log scale. If so, check your values again.

3.2.4.   DC Amplifiers

The development of electronics over the past few years has made modern circuitry extremely stable and reliable, eliminating much of the need for constant readjustment and calibration. AC amplifiers in particular do not normally need any form of operator adjustment. However, *DC amplifiers* do have a special problem which is worth discussing, because of the effect that it has on the design and operation of many instruments. The problem arises because the characteristics of the electronic components of a circuit do not remain constant but gradually change. This may be due to an ageing process, fluctuations in temperature or some other long-term effect. These results in a slow *drift* in the DC voltages inside the amplifier.

How will this affect the relative performances of an AC and a DC amplifier? We can consider the drift in voltage mentioned in the previous paragraph to be equivalent to a low frequency AC signal. However, the distinction between a drift signal and a normal AC signal is that the effective period of oscillation of the drift signal is much longer than the period of our AC signals. Thus, in our Fourier transform, the drift signal has a very low frequency indeed (see Fig. 3.2d), and this will normally be less than the low-frequency cut-off point of an AC amplifier. Since drift signals fall outside the bandwidth of the AC amplifiers, the output of the AC amplifier will not be directly affected by changes due to fluctuation of temperature or ageing of components.

The output of a DC amplifier, however, *will* gradually change in response to the low-frequency drift variations of voltages inside the amplifier. The observed effect of this is that, even if a constant volt-

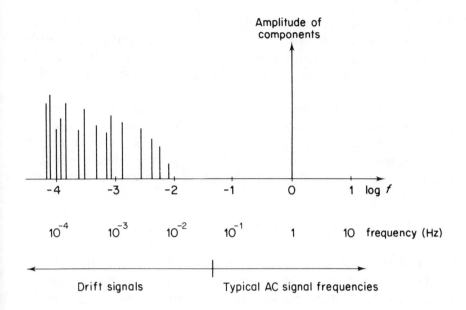

Fig. 3.2d. *'Drift' as Fourier components*

age is applied to the input of a DC amplifier, the *output voltage* will gradually drift with time. The drift may be slow – over a period of hours, days, or years.

We can correct this drift by allowing the operator to add a Zero Offset Voltage, V_{os}, to the input which cancels out (temporarily) the effect of the drift voltage. This control is usually set by first short-circuiting the input to the amplifier so that the input voltage is zero, and then adjusting the Offset (zero) Control until the output voltage is also zero. The frequency of the need to re-set the 'zero' depends on the sensitivity of the amplifier and its rate of drift.

We shall see in the following exercises that sometimes we can quantify the drift at the output of an amplifier by looking at the input to the amplifier and asking the question: 'What offset voltage, V_{os}, must be added to the input to counteract the effects of drift?' For an ideal amplifier the answer should be that the offset voltage should be zero, $V_{os} = 0$.

Π   If a DC amplifier (possibly in a chart recorder) has a maximum drift rate of 0.01 mV min^{-1} at the input to the amplifier, calculate the maximum drift error in the output over a period of 60 min, assuming the amplifier has a gain of 100.

The drift-error voltage has been specified at the input of the amplifier, so we must multiply by the gain to find the output voltage. We must also realise that drift voltage was quoted as a rate so to calculate an actual error voltage we must multiply by the appropriate length of time, viz. 60 min. Thus the maximum output-error voltage would be

$$\pm\ 0.01\ \times\ 60\ \times\ 100\ =\ \pm\ 60\,\text{mV}.$$

Note that this should represent an upper limit in the drift, and very often the actual drift would be much less. Notice also that we cannot predict which way the drift will go, hence the use of the $\pm$ sign.

It may help to make the problem more realistic if you realise that a pH meter is essentially a DC amplifier. When you adjust the 'Set Buffer' control on a pH meter you are in fact making a 'zero-offset' correction in a DC amplifier which is amplifying the DC signal from the pH electrode. This offset control allows you to correct for the variations in DC Offset between different electrodes, and for drifts that occur in the behaviour of the pH-meter system itself. You should always recheck your buffer settings at regular intervals to eliminate the effect of DC drift.

3.2.5. Response Time of DC Amplifiers

Strictly speaking a DC signal is one of *zero* frequency ie something that *never* changes. However, in reality, any measurement we ever make involves some change. We may, for example, measure the difference in pH between a buffer solution and an unknown solution. Alternatively the signal that we are measuring may itself be changing with time. If we are dealing with a signal that changes with time, then we must consider the Fourier Transform of that signal, see Section 3.1.6. We must make sure that the band-width of the amplifier is sufficiently large to cope with the band-width of the analytical signal. We have already introduced this idea in Section 3.1.6, and in SAQ 3.1b. It is often convenient to express the band-width of the amplifier in terms of a *time*, instead of frequency, Δf, by using Eq. 3.9.

$$\tau \simeq 1/(k \times \Delta f) \tag{3.9}$$

This is called the *Response Time* or *Time Constant* of the amplifier, and gives an indication of the *minimum* time taken by the amplifier to respond to a sudden change in its input. The value of the constant k depends on the exact definition used for tau.

If τ is the Response-Time then $k = 2.7$.

If τ is the Time-Constant then $k = 2\pi$.

The different ways of defining response time by using different values of the constant, k , are discussed elsewhere.

∏   If the response time of an amplifier is halved what happens to its band-width?

The relationship between time and frequency is always in the form of a reciprocal. Thus *halving* the time must *double* the frequency band-width, and this will be independent of the particular value of k used.

We shall see later that the same calculations apply to a whole instrument as well as to a single amplifier, and we are then interested in how quickly the instrument responds to changes in an analytical signal.

3.2.6.   Buffer Amplifiers and Power Amplifiers

Not all amplifiers increase the *voltage* of a signal. The primary purpose of both *buffer* and *power* amplifiers is to increase the *current* in a signal.

We saw in Section 2.3.6 that the ideal measurement of a voltage source can be made only if zero current is taken from the source. With a pH-meter for example, if a current of more than a few nanoamps (1×10^{-9} A) is taken from the electrode, then the pH reading will be wrong.

A modern *buffer* amplifier requires an input current only of the order of a picoamp (1×10^{-12} A), and can give an output current of milliamps. There is normally no voltage gain: the output voltage is equal to the input voltage. This allows the voltage from a pH-electrode, for example, to be recorded accurately, and then transferred with sufficient current (and hence power) to operate either a display meter or further electronic circuitry.

We have just seen that a buffer amplifier converts a near-zero current into milliamps. A *power* amplifier, however, converts a current of milliamps into a current of amps. A power amplifier has to be used when a voltage signal is needed to drive a component which requires considerable power (volts × amps), eg a loudspeaker, a lamp, a heater, or a motor. As for the buffer amplifier, the power amplifier does not usually change the voltage level of the signal.

Referring back to Section 2.3.6 we see that the buffer amplifier has a *very high input-impedance* and the power amplifier has a *very low output-impedance*.

3.2.7. Logarithmic Amplifiers

In the normal amplifier the output voltage, V_0, is a *linear* function of the input voltage V_i.

$$V_0 = G \times V_i \qquad (3.10)$$

However you may meet other types of amplifier. In the *Log Amplifier* for example, the output voltage V_0 is given by Eq. 3.11.

$$V_0 = G \times \log(V_i) \qquad (3.11)$$

This amplifier has a particular application in spectroscopy in displaying the value of *absorbance* which is a logarithmic function of the intensity of light passing through the sample.

SAQ 3.2a

A particular type of small transistor radio has an output-amplifier system with a frequency response from 150 Hz to 6.00 kHz.

(*i*)   Is this an AC or a DC amplifier?

(*ii*) Calculate the band-width of the amplifier.

(*iii*) Explain why the small band-width gives poor quality sound.

SAQ 3.2b

A block diagram of a simple pH meter is given in Fig. 3.2e. Two amplifiers A and B are shown.

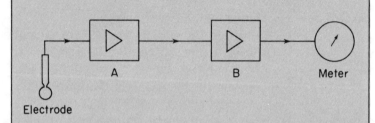

Fig. 3.2e

The input to A from the pH electrode is a voltage signal which changes by 59.0 mV for unit change on the pH scale. The output from A also changes by 59.0 mV per pH unit.

The output from B operates the display meter and is a voltage signal which changes by 1.00 volt for unit change on the pH scale.

(*i*) State which type of amplifier should be used (*a*) for A, and (*b*) for B.

Choose from: AC amplifier, DC amplifier, Buffer amplifier, Power amplifier, Logarithmic amplifier.

(*ii*) Based on the information in the text, estimate the voltage gain for each of the amplifiers A and B.

SAQ 3.2b

3.3. FILTERS

3.3.1. Introduction

Amplifiers are not the only type of circuit that is used to process signals. There is a whole range of signal-processing circuits that have effects on the signal other than amplification of the voltage or current. In this part we discuss one other important type of circuit, the *filter*.

3.3.2. The Electronic Filter

Let us first consider the variation of gain with frequency which is shown in Fig. 3.3a, and decide on the function of the circuit which has those characteristics. Certainly it is not an amplifier because the gain never exceeds unity.

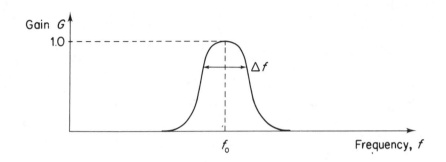

Fig. 3.3a. *Bandwidth, Δf, of a filter*

However we can see that the gain for all frequencies except those near f_0 is virtually zero. This will mean that for frequencies near f_0 the signals will pass through the circuit without attenuation, but that all other frequencies (outside the range Δf) will be severely attenuated and absorbed. This is a circuit that will let through certain frequencies (f_0) and block all others – a *frequency filter*. It is possible to assign a *band-width*, Δf, to an electronic filter in the same way as we did for an amplifier in Section 3.2.3. We can make a very close comparison with an optical filter such as green glass. This glass will let through wavelengths (or frequencies) that correspond to green light but it will block all other wavelengths (or frequencies), red, blue, yellow.

What is important about an electronic filter is the effect that it will have on a signal. Consider, for example, a square wave which has a Fourier Transform as in Fig. 3.3b(i), and which then passes through a filter with a gain as in Fig. 3.3b(ii). Only those Fourier components of the signal that lie within the band-width of the filter will pass through the filter. Thus the Fourier Transform of the signal which emerges from the filter will be given by Fig. 3.3b(iii). The effect of this filter on the waveform is illustrated in Fig. 3.3c. Note that because only one frequency component is allowed through, the final waveform is that of a single sinusoidal component.

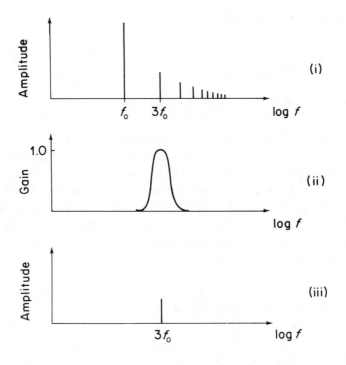

Fig. 3.3b. *Effect of a filter on Fourier components*

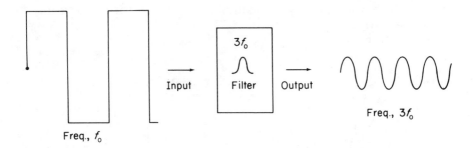

Fig. 3.3c. *Effect of a filter on waveform*

3.3.3. Different Types of Filter

We indicate below the significance of particular types of filter when applied to certain signals.

Tuned filter. The Gain of this type of filter is given in Fig. 3.3b(*ii*).

You can see that the band-width is very small and that the filter will virtually let through only a single frequency component of a signal. The exact frequency is chosen by tuning the filter, hence the name. The example chosen above to illustrate the effect of the tuned filter was that of the square wave, and we assumed that the filter frequency, f_o, was 3 times the fundamental frequency in the square wave. We could also have chosen a filter tuned to any of the other frequency components.

Notch filter. This is virtually a mirror image of the previous filter in that only one particular frequency is blocked.

One of the problems of interference in the electronics of instruments is the 'pick-up' of a 50 Hz signal due to the mains supply in the laboratory. Fig. 3.3d(*i*) illustrates the Fourier Transform of an analytical signal with a superimposed mains signal. After passing through a notch filter tuned to 50 Hz, the mains interference is eliminated, Fig. 3.3d(*ii*). The component at 50 Hz in the analytical signal is also lost, but on balance, a small amount of degradation of the analytical signal may be a small price to pay for eliminating the mains interference.

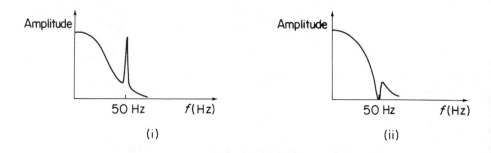

Fig. 3.3d.   *Rejection of 50 MHz mains pick-up*

Low-pass filter. The origin of the name of this filter should be obvious from the gain characteristics (see Fig. 3.3e) only frequencies below f_h are allowed to pass. This also includes DC signals. We shall see in a later section how this is important in limiting the noise in the output of an instrument, and in fact how many 'noise-limiting' controls on instruments use a low-pass filter.

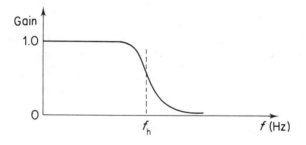

Fig. 3.3e. *Low-pass filter*

The Gain characteristics of *high-pass* and *band-pass* filters are shown in Fig. 3.3f. However, given the name of the filter, you should by now have been able to guess the shape of the Gain characteristics.

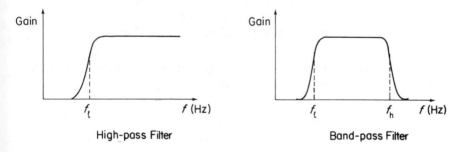

High-pass Filter   Band-pass Filter

Fig. 3.3f. *High-pass and band-pass filters*

Π   To answer the following questions you need to refer back to the waveform introduced in Fig. 3.1f. This waveform consists of a series of pulses alternating in amplitude between V_R and V_s.

Assume that this waveform, W, is passed separately through two tuned filters A and B of frequencies 50 Hz and 25 Hz respectively, as in Fig. 3.3g.

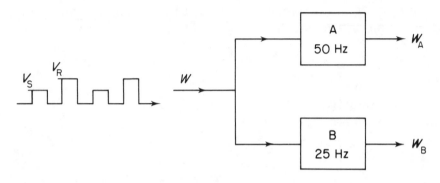

A and B are tuned filters

Fig. 3.3g

(*i*)   Describe the two waveforms, W_A and W_B, that will emerge from filters A and B. Will they be square waves or sine waves and what will be their frequencies?

(*ii*) The amplitudes of each of these components from A and B will tell us something different about the magnitude of V_R and V_S. What information do they carry?

You may need to study the exercises in Section 3.1.4 again together with Fig. 3.1g before attempting this problem.

The first step is to look at the Fourier Transform of the waveform. We do not have all of this transform available but we do know from Fig. 3.1g that there are two frequency components: at 50 Hz and 25 Hz.

(*i*)   The filter 'A' will pass only the 50 Hz component, and filter 'B' will pass only the 25 Hz component.

The frequency components in a Fourier Transform are each sinu-soidal waves, so that if just *one* component is isolated as in this case for each of A and B, then it will appear as a *sinusoidal* wave.

(*ii*) Again looking at the solution to the exercises in Section 3.1.4 we see that the amplitude of the 50 Hz signal passing through A is related to the average value of V_R and V_S ie $(V_R + V_S)/2$. We can also see that the 25 Hz signal passing through B is related to the difference between V_R and V_S ie $(V_R - V_S)$. Thus our signal-processing elements have sifted out different pieces of information from our original signal.

It may be that the two pulses, V_R and V_S, correspond to the intensity of light detected from the two beams of a double-beam spectropho-tometer. The output from A then gives an indication of the total intensity of light reaching the detectors whereas the output from B gives an indication of any imbalance between the two beams.

3.3.4. Time-constant of a Filter

We continue here to develop a little further the concept of time-constant already introduced in Section 3.2.5. We saw that the idea of a time-constant or response-time is applied to an amplifying system which has a response down to zero frequency (DC). In terms of the filters that we have introduced this applies only to the low-pass filter.

It is common practice to use the term *Time-Constant*, in reference to a *Low-Pass* filter, defined by Eq. 3.12,

$$\tau = 1/(2\pi \Delta f) = 1/(2\pi f_h) \tag{3.12}$$

where f_h is the high frequency cut-off point of the filter. (The band-width of a low-pass filter is equal to f_h). We shall see below how this type of filter can be used to eliminate unwanted high-frequency noise, while allowing a low-frequency signal to pass through.

∏ The Fourier transforms of two signals S and N are given in Fig. 3.3h. Each of the signals is passed through a low-pass filter with a time constant of 0.1 s. Sketch the Fourier transform of each signal after it has emerged from the filter.

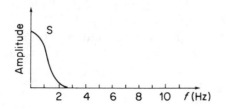
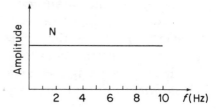

Fig. 3.3h

The first step is to convert the time-constant of the filter into a high-frequency cut-off, f_h. This is done by using Eq. 3.12 which gives

$$f_h = 1/2\pi\tau = 1.6 \text{ Hz}$$

As this is a low-pass filter it will pass only components with frequencies below f_h. We can see immediately that most of the components of signal S have frequencies below f_h and thus this signal will be largely unaffected. Its Fourier Transform will remain the same. The signal N, however, has frequency components over the complete spectrum. Those components with frequencies much above f_h will be completely absorbed, giving a resultant transform as in Fig. 3.3i.

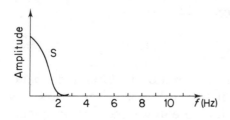
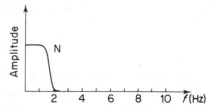

Fig. 3.3i

We have exemplified in this question the action of a low-pass filter allowing through an analytical signal S, while removing as many high-frequency components as possible from the unwanted noise signal hearing spectrum, N.

3.3.5. Conclusion

In this section we have assumed that the maximum gain for a filter would be unity. This has allowed us to consider the properties of the filter separately from those of the amplifier. It is common practice to combine the two so that one circuit acts both as a filter and as an amplifier. This idea should not present you with any new problems.

Most filter circuits also affect the *phase* of any signal components which have frequencies near the cut-off points. This can lead to some very complicated effects, but it is beyond the scope of this text to attempt a discussion on this topic, and we have ignored the effects in our simplified treatment.

SAQ 3.3a   A signal as in Fig. 3.3j consists of a DC voltage and an AC 'ripple'. This type of voltage may exist in the DC power supply for an instrument, see Section 2.2.4. If we wish to measure the amplitude of the AC ripple, V_{ac}, we may find problems because of the large DC voltage associated with it.

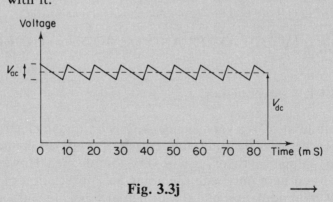

Fig. 3.3j $\longrightarrow$

SAQ 3.3a
(cont.)

We can however pass this signal through a filter so that we have only the AC parts of it. Describe the characteristics of a filter which will do
this. Estimate the value of the high or low cut-
off frequency or band-width as appropriate for
the type of filter you have chosen.

3.4. MODULATION AND DEMODULATION OF SIGNALS

3.4.1. Introduction

Modulation is a very widely used technique in electronics and signal
processing. The different methods of modulation and the associated
mathematical theory can be very complicated. However, we are concerned here only with obtaining an appreciation of the reasons for
using modulation and the principal ways in which it affects the design of instruments.

For many different reasons it is frequently inconvenient or inefficient to handle an analytical signal in its 'normal' state. As a simple example, we start by assuming that the analytical signal is in the form of a DC voltage. We have already seen that any DC signal can be subject to errors due to drift. If, for example, we amplify this DC voltage through a DC amplifier then we also amplify the error due to drift. If we could use only an *AC* amplifier to amplify the DC signal, then we could eliminate the problem due to drift. This may seen to be a contradiction in terms and an unlikely possibility but in fact it is exactly what we can do when we use a *modulated* DC signal.

3.4.2. Amplitude Modulation of a DC signal

A very simple way to modulate a DC signal is to switch it alternately on and off at a modulation frequency, f_m. This then converts the original DC signal into a square wave of frequency, f_m. A system for doing this is illustrated in Fig. 3.4a(i). The DC voltage, V_{dc}, is applied at the input on the left, and a switch, S, is made to oscillate at a frequency, f_m, between contacts A and B. When S is at A, the voltage, V_{dc}, is connected to the next stage so that $V^1 = V_{dc}$. When S is at B, the next stage is connected to earth (0 *V*) so that $V^1 = 0$.

The DC signal is thus converted into a square wave, V^1, with a frequency, f_m, see Fig. 3.4a(ii).

However, this wave still contains a DC component, which can be removed by passing the signal through a high-pass filter whose low-frequency cut-off is much less than the modulation frequency, f_m. The result is a square wave, V_{ac}, see Fig. 3.4a(iii). The position, or phase, of the switch, S, is shown in Fig. 3.4a(iv).

The Fourier Transform of the signal before and after modulation is shown in Fig. 3.4b. The FT for the square wave is derived from the FT of the square wave in Section 3.1.4.

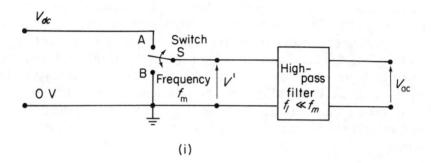

(i)

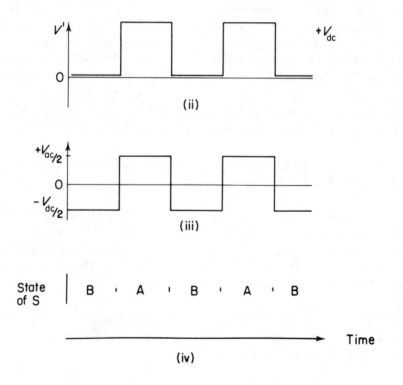

Fig. 3.4a.   *Modulation of DC signal*

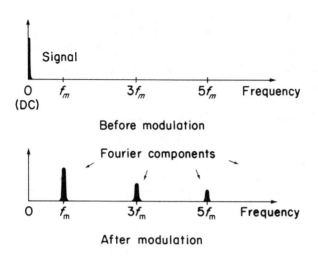

Fig. 3.4b.   *Square wave modulation of DC signal*

The information that was contained in the original signal is the *amplitude*, V_{dc}, and the *polarity*, $+/-$, of the DC voltage. Have we lost any of this information in modulating the signal? To find the answer to this question work through the following exercise.

∏   If the magnitude of the DC signal in Fig. 3.4a(i) is 0.6V calculate the peak-to-peak voltage of the AC signal, V_{ac}. If you are not sure what is meant by peak-to-peak voltage then refer back to Section 2.1.9.

The peak-to-peak voltage of the AC signal is the amplitude from crest to trough, which in Fig. 3.4a(iii) is equal to $+V_{dc}/2$ $-$ $(-V_{dc}/2)$ $=$ V_{dc}. The answer to the question is 0.6 V, since V_{dc} is the magnitude of the DC signal. The important point to note is that the amplitude of the modulated signal depends on the magnitude of the un-modulated signal.

From this exercise it should be clear that the amplitude of the modulated signal, V_{ac}, is directly related to the magnitude of the unmodulated signal V_{dc}. Thus the information contained in the *magnitude* of V_{dc} is transferred to the *amplitude* of V_{ac}.

∏   Re-draw the diagrams in Fig. 3.4a(*ii*), (*iii*) and (*iv*) with V_{dc}
     a *negative* voltage instead of the positive voltage used in the
     original diagram. Include V^1, V_{ac} and the phase of S, and
     take care to ensure that the timing is the same on each of
     them.

The correct answer is shown in Fig. 3.4c.

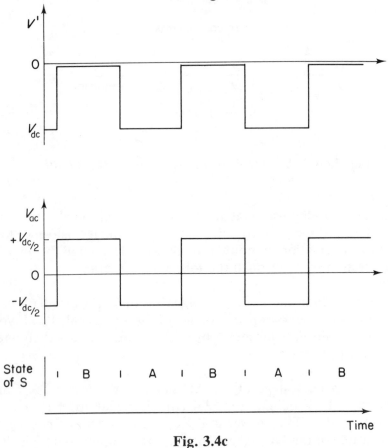

Fig. 3.4c

V^1 must go from 0 V to V_{dc} (which has a negative value) when S
switches from B to A. This is because V^1 is connected directly to
V_{dc} whenever the switch goes to A. The phase of V_{ac} should be
exactly the same as that of V^1. The only difference is that the DC
component has been removed by the filter.

∏ By comparing Fig. 3.4c with Fig. 3.4a, can you explain what effect the polarity of the DC voltage, V_{dc}, has on the AC signal, V_{ac}?

The only difference between V_{ac} for positive and negative values of V_{dc} is one of *phase*. If this is not clear, compare the displacement of V_{ac} in the diagrams when S is switched to A, for example. With *S* at A, V_{ac} is positive when V_{dc} is positive, and when V_{dc} is negative V_{ac} becomes negative. The *polarity* of V_{dc} is reflected in the *phase* of V_{ac}.

It is important to note that no information has been lost and that the same information is now being carried in an *AC signal* instead of a *DC signal*. The *magnitude* of the DC voltage is represented by the *amplitude* of the AC signal. The *polarity* of the DC voltage is represented by the *phase* of the AC signal.

This means that we can now use an AC amplifier instead of a DC amplifier, and immediately we have reduced the problems caused by DC drift mentioned in Section 3.2.4.

This specific method of modulating a DC signal by switching on and off is generally referred to as *chopping*. The point at which the signal is 'chopped' is not necessarily in the electronic circuit. For a spectrophotometer, the actual light beam may be 'chopped' by rotating mirror even before it has passed through the sample and before it has been converted by a photo-cell into an electric current.

3.4.3. Demodulation

We have described above a process whereby, through modulation, an analytical signal is converted into some form more convenient for processing, but at some point it is probable that we would wish to bring the information back to a form similar to the original. This reverse process is called *demodulation*.

One method we could use for demodulation has already been introduced in a slightly different context in Section 2.2.4. In that section we were considering how we could convert AC mains power into DC power. Although we are not now primarily concerned with power, we do want to convert an AC signal at a frequency f_m into a DC signal. To do that we can again employ the *rectifying* properties of a diode circuit followed by the smoothing properties of the capacitor, see Fig. 3.4d. The amplitude of the DC signal, V_{dc}, produced by this circuit is directly proportional to the amplitude of the input signal, V_{ac}.

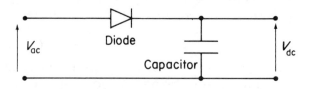

Fig. 3.4d.   *Demodulation by rectification*

3.4.4.   Phase-sensitive Detection

The simple rectifying method described above does not take account of the *phase* of the AC signal. As is implied in Fig. 3.4d, the DC output from the circuit will always be positive irrespective of the phase of the AC signal, so that any information that was carried in the phase of the AC signal has now been lost.

If we wish to retain information then we must use a Phase-Sensitive Detector (PSD) circuit or a Synchronous Rectifier circuit. To see what is necessary for phase-sensitive detection, consider for a moment how the phase of V_{ac} in the previous exercises changes for positive and negative values of V_{dc}. We are able to identify the phase of V_{ac} by comparing it with the timing sequence of the positions A and B for switch S.

The differences in phase of a wave can be detected only by comparison with some timing reference-signal, hence the sequence of the positions of S. Thus we can say that a PSD system must compare the modulated signal with a *reference timing-signal*. This was introduced in Section 2.2.7 in which it was shown that it was necessary to have a *reference wave* in addition to a *signal wave*.

Π In the circuit in Fig. 3.4e(i) a square wave, V_{ac}, is applied at
the input on the left. The switch is opened (0 – no contact)
and closed (C – electrical contact) at the same frequency as
the incoming wave.

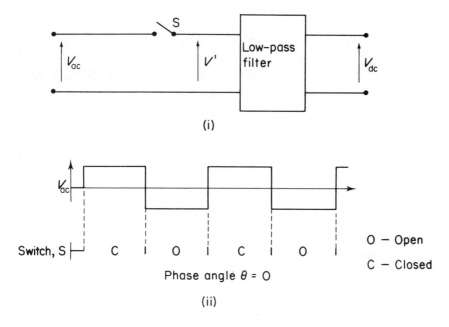

Fig. 3.4e

Assume that the switch is *in phase* with V_{ac} if S is closed (C)
at exactly the same time as V_{ac} is positive, see Fig. 3.4e(ii).
In this state the phase angle (θ) between V_{ac} and S is zero
($\theta = 0$). The low-pass filter has the effect of averaging any
variation in the voltage, V^1, to give a constant DC voltage,
V_{dc}.

If the DC voltage, V_{dc}, is $+3$ V when S is in phase with V_{ac}
($\theta = 0$), then calculate V_{dc} when

(i) S is out of phase with V_{ac}, $\theta = 180°$,

(ii) The phase-angle between S and V_{ac} is $90°$, $\theta = 90°$.

If the switch is *in-phase* with the signal, it is closed when V_{ac} is positive and passes a positive voltage through to the output, V_{dc}.

(*i*)   If S is *out-of-phase* with the signal, the switch is closed when V_{ac} is negative. Hence a negative voltage is passed through to the output, giving $V_{dc} = -3$ V.

(*ii*) If the phase of S is 90° away from that of the signal, the switch S will pass equal amounts of positive and negative voltage. There will thus be no net DC component, and the low-pass filter circuit will give an average value of zero at the output, $V_{dc} = 0$ V.

If this is not clear, then look at Fig. 3.4f which gives the relative phases between S and V_{ac}. You should also refer back to Section 2.2.7 and SAQ 2.2d.

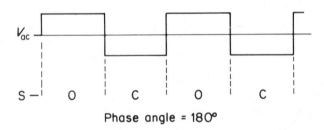

Phase angle = 180°

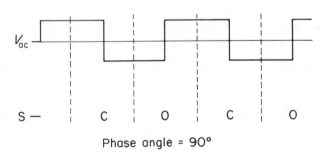

Phase angle = 90°

Fig. 3.4f

3.4.5. General Modulation Methods

In this section we have discussed only one of many different forms of modulation. In general the signal is combined with a constant signal of higher frequency – the modulation frequency, f_m. This high frequency signal is often called the *carrier wave*. The information that was contained in the original analytical signal is now contained in a new signal at a frequency at or near the modulation frequency.

There are different methods of combining the analytical signal and the modulation frequency, but the three principle methods result in the information being carried as the *amplitude* or the *frequency* or the *phase* of a *carrier* wave. You may remember the terms AM and FM used on radio receivers. They stand for amplitude modulation and frequency modulation respectively, and correspond to different ways of encoding the audio frequency signal onto a much higher radio frequency wave for transmission.

SAQ 3.4a | In the diagram of an optical detection system in Fig. 3.4g, the light beam is 'chopped' by a disc rotating at 50 rotations per second. The disc has 4 segments as shown: two segments block the light completely and two transmit the light.

The intensity of light is converted into an electrical signal by a photo-detector. This signal is to be amplified by a tuned amplifier.

At what frequency should the amplifier be tuned?

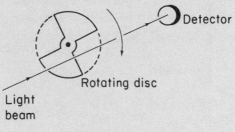

Fig. 3.4g

SAQ 3.4a

SAQ 3.4b   A block diagram, Fig. 3.4h, represents a system which amplifies a DC signal by using a modulation technique, followed by phase-sensitive detection. However, too many information pathways have been drawn in the diagram. Only *one* of the paths, A, B, or C, should be kept in the diagram. Which *one* pathway is essential? Choose from A, B, and C.

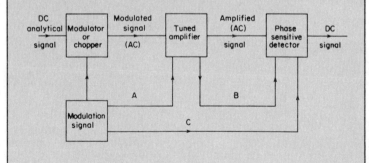

Fig. 3.4h

SAQ 3.4b

3.5. NOISE IN INSTRUMENTS

3.5.1. Introduction

Noise is one of those terms which have meaning both in everyday usage and in a scientific context. Broadly speaking the two meanings are similar, but, we must be careful that we know exactly what we mean when we talk of 'noise' in an instrumental measurement. Noise is generally assumed to be an unwanted phenomenon. One may feel that, with enough effort, *everyday* noise can be eliminated (eg moving to the country!). However, there is a certain amount of noise in every scientific *measurement* that *cannot* be reduced. The accuracy of every measurement is ultimately limited by noise. Some instrumental noise can be reduced by an improvement in equipment design or experimental procedure. However we see that nature itself is a fundamental source of noise, which just cannot be 'switched-off'.

3.5.2.   Analytical Signals and Noise Signals

Every *analytical signal* has, mixed with it, a certain amount of *noise signal*.

Intuitively we believe that we can distinguish between something we call a 'signal' and something we call a 'noise'. In Fig. 3.5a, there is a graph of an electrical signal (*i*) and electrical noise (*ii*). We expect the signal to have a certain 'smoothness' or 'regularity' about it, and the 'noise' to have a random character.

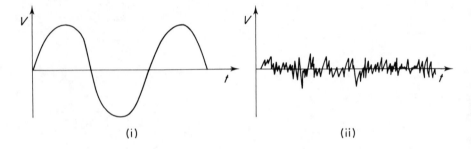

(i)   (ii)

Fig. 3.5a.   *Signal and noise*

Even if we add these two together – signal **+** noise, as in Fig. 3.5b, it is still easy to see that there is a signal present. In this case, where the signal is obvious, we say there is a large *Signal-to-Noise Ratio*.

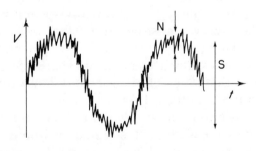

Fig. 3.5b.   *Signal and noise*

If S is the root-mean-square (rms) amplitude (see Section 2.1.9) of the signal, and N is the rms amplitude of the noise, the signal-to-noise ratio is then defined below.

$$\text{Signal-to-Noise Ratio} = S/N$$

In the example given in Fig. 3.5b, S/N has a value of about 5, and the signal is clearly visible. What happens if the amplitude of the signal is less than that of the noise, for example if $S/N = 0.5$? Judge for yourself by looking at Fig. 3.5c. You can see that it is very difficult, by eye, to decide whether or not there is any signal buried in the noise; a measurement of the magnitude of the signal would be even more difficult.

Fig. 3.5c. *Noise and signal (?)*

The above examples bring out two very important points about signals and noise.

(*a*) As far as the electronic circuits are concerned there is *no* difference between analytical signals and noise signals: they are both a variation of voltage with time. Just as it is possible to produce a Fourier Transform of an analytical signal, it is also possible to draw out the Fourier Transform of a noise signal. A Fourier component of a *noise signal* at, say, 1 kHz will be amplified by an amplifier in exactly the same way as a 1 kHz Fourier component of an *analytical signal*. It is not possible to design an amplifier (or other circuit) which will reject one but not the other.

(*b*) The way in which we recognise the difference between a noise signal and an analytical signal is that we expect the analytical signal to have some particular *pattern* of behaviour that the noise signal does not have. For example we would expect that if we repeat the experiment the analytical signal would be almost exactly the same, whereas the noise signal should change in a completely random way. This method of repetitive measurement is used extensively in data handling systems.

Provided that we remember the two points given above, it is more convenient in the rest of this section to continue to use the rather inexact terms 'signal' to mean an analytical signal, and 'noise' to mean a noise signal.

3.5.3.   Sources of Electronic Noise

There are four main sources of noise in electronic circuits:

— Johnson noise,

— Shot noise,

— Flicker $(1/f)$ noise,

— Enviromental noise.

We describe below the origins of each of these types of noise. Fig. 3.5d gives typical Fourier Transforms (or frequency spectra) for a combination of all these types of noise.

Johnson noise arises because the electrons which carry an electric current always have a thermal motion (see Section 2.1.2). This random motion increases as temperature increases and causes very small fluctuating voltages to appear across any resistance in the electrical circuit. It is possible to amplify these noise voltages, and if they are fed through a loudspeaker a constant 'hiss' can be heard. There is an analogy in sound, in that if a conical sea shell is used as an amplifying horn by placing it close to the ear, then a similar 'hiss' can be heard. This is, in fact, the amplified noise of the random motion of the molecules of the air in the shell. If a Fourier Transform of Johnson noise is drawn then it can be seen that it contains compo-

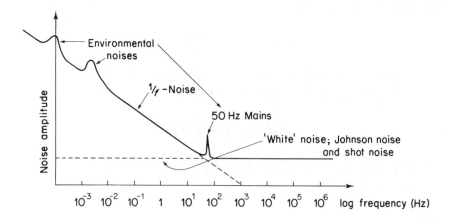

Fig. 3.5d. *Noise components*

nents at all frequency values, cf Fig. 3.5d. It is said to be *white* noise. This is a comparison with white light, which contains all the possible frequency components in the visible spectrum. The rms voltage due to Johnson noise across a resistance, R ohms, is given by, Eq. 3.13.

$$V_{rms} = \sqrt{4\,kTR\Delta f} \tag{3.13}$$

where k is Boltzmann's constant $= 1.38 \times 10^{-23}$ J K^{-1}, T is the Absolute temperature in Kelvins, and Δf is the band-width of the signal-processing circuit.

∏   An amplifier has a gain, G, of 100 and a band-width from 50 Hz to 100 kHz. The amplitude of Johnson noise at the output of the amplifier is 0.300 mV at room temperature (295 K).

Calculate:

(*i*)   the new noise signal if the band-width was reduced so that it extended only from 990 Hz to 1010 Hz,

(*ii*)   the noise signal if the temperature was *also* reduced from room temperature to 77.0 K.

We must use Eq. (3.13).

$$V_{rms} = \sqrt{4\,kTR\Delta f}$$

In this question k and R are constants so that we can re-write the equation with a new constant, K' as follows.

$$V_{rms} = \sqrt{K'T\Delta f}$$

Provided that the gain of the amplifier is constant it will not enter into our calculations.

Thus we have:

$$0.300 \times 10^{-3} = \sqrt{K' \times 295 \times (100 \times 10^3 - 50)}$$

$$= \sqrt{K' \times 295 \times 100 \times 10^3}$$

(Within the accuracy of the figures given.)

$$\therefore \quad K' \quad = 3.05 \times 10^{-15} \; V^2 \; K^{-1} \; Hz^{-1}$$

If you have calculated 50 Hz (= 100–50) for the band-width, then you did not notice that the high-frequency cut-off, f_h, is expressed in Kilohertz.

(*i*)   Here the band-width is 1010–990 = 20.0 Hz.

$$V_{rms} = \sqrt{3.05 \times 10^{-15} \times 295 \times 20.0}$$

$$= 4.24 \times 10^{-6} = 4.24 \; \mu V$$

(*ii*)   The temperature is now also reduced to 77.0 K

$$V_{rms} = \sqrt{3.05 \times 10^{-15} \times 77.0 \times 20.0}$$

$$= 2.17 \times 10^{-6} = 2.17 \; \mu V$$

You can now see that it is possible to make a dramatic reduction in noise by using a narrow-band amplifier instead of a broad-band amplifier. A reduction in temperature (by using liquid nitrogen) can also reduce noise. Since much of the noise in a detection system originates in the detector itself it is a fairly common occurrence for the detector to be cooled to a temperature below ambient, eg the detector in an energy dispersive X-ray fluorescence system.

Shot noise occurs because the flow of electric current is not entirely smooth and continuous, but consists of a random flow of electrons each carrying a fixed charge, e ($= 1.6 \times 10^{-19}$ coulombs). This discrete nature of electrical charge causes very slight fluctuations in the actual flow of charge. Consequently shot noise is expressed in terms of a *current noise* with an rms value given by Eq. 3.14.

$$I_{rms} = \sqrt{2\,Ie\Delta f} \qquad (3.14)$$

Note that this depends on the magnitude of the total current, I. Just as for Johnson noise the actual noise level is proportional to the square root of the band-width, Δf, of the amplifying system.

∏   Can Shot noise be called white noise?

We already know that Johnson noise is 'white' noise. The amplitude of the components is the same for all frequencies. You can see this in the expression for V_{rms} in Johnson noise: it depends on the band-width, Δf, *not* on the value of the frequency, f. The value of V_{rms} for shot noise has the same dependence on Δf as for Johnson noise. The amplitude is independent of the actual value of the frequency, f. It is white noise.

Flicker (or one-over-f) noise. The first comment to make about flicker noise is that it is definitely *not white* noise, see Fig. 3.5d.

There are many different sources of this type of noise, some of which are not well understood. What is certain, however, is that it is always present and dominates all other types of noise at frequencies below about 100 Hz. The principal characteristic of this noise, which also gives it one of its names, is that the amplitude of the noise is inversely proportional to the frequency, f.

$$V_{rms} \propto 1/f$$

This is shown in Fig. 3.5d, where a plot of the variation of V_{rms}, at low frequencies, against log f gives a straight-line graph with a negative slope. Obviously at increasing frequencies, Flicker noise will decrease, eventually becoming smaller than Johnson noise. Flicker noise will on the other hand become dominant at very low frequencies in the noise spectrum.

It was mentioned in Section 3.2.4 that the 'zero' of a DC amplifier changes slowly with time. This is just one example of Flicker or one-over-f noise at a very low frequency. If, for example, this change occurs over a period of a day, $T = 24 \times 60 \times 60$ s, this is equivalent to a frequency of, $f = 1/T = 1.16 \times 10^{-5}$ Hz. It may seem odd to you to talk of such a low frequency, but variations of time equal to a day, month or year are indeed *cyclic* even if they do not show a smooth sinusoidal oscillation.

Environmental noise. There are very many different noises to be included in this category. The most common of these is a result of the mains electricity circulating around the walls of the laboratory and buildings in which we work. Because of the electric and magnetic fields created by these mains currents, an AC signal of the same frequency (50 Hz in the UK) can be induced into the wiring of the instrument. This is referred to as *mains pick-up*. It is quite common for an instrument to have a notch filter included specifically to cut out any frequency component at mains frequency, see Section 3.3.3.

Other sources of environmental noise include transient electrical pulses that occur on the mains supply due to the sudden switching of heavy machinery such as lift motors in the building. The daily variations of temperature in the laboratory, or the intermittent effect of direct sunlight, may induce slight changes in the performance of certain components. This does indeed represent a 'noise' signal even though it is of very low frequency. You can probably see that most environmental noise has a low frequency. Indeed it would also be possible to include many of these sources of noise under the Flicker $(1/f)$ Noise category.

3.5.4. Reduction of Noise

Many different methods have been used to reduce noise levels and increase the S/N ratio. However, we mention here only three of the main categories, although other methods will be introduced elsewhere. We introduce two of these methods in the following exercises.

∏   For this question use the two signals S and N in Fig. 3.3h. When these two signals pass through a particular amplifying system with a band-width of 10 Hz the ratio of the amplitudes of S and N is given by:

$$S/N = 5.0.$$

Calculate the new S/N ratio when a low-pass filter of time-constant 0.1 s is included in the amplifying system.

Look back at Section 3.3.4 if you are unsure about the effect of a filter.

In answering this question you should be careful not to confuse *time-constant* and *band-width* for a low-pass filter.

If $\tau = 0.10$ s

$$f_h = 1/2\pi\tau = 1.6 \text{ Hz}$$

Since this is a low-pass filter, f_h, also equals the band-width,

$$\Delta f = 1.6 \text{ Hz}$$

The filter reduces the band-width by a factor of $10/1.6 = 6.25$.

The noise level will be reduced in proportion to the *square root* of this factor, ie by $\sqrt{6.25} = 2.5$. This is because the noise signal, V_{rms}, is proportional to the square root of the band-width.

If the noise is reduced by a factor of 2.5, the S/N ratio is *increased* by a factor of 2.5.

$$\text{Resultant } S/N = 5.0 \times 2.5 = 12.5.$$

The situation, described, may occur, for example, in the output from a recording spectrophotometer. A low-pass filter is often included in the output of an instrument so that the operator can eliminate high-frequency noise components. This may be referred to as a Noise Limiting System. The amount of noise suppression is increased by increasing the time-constant of the filter. You must be careful to ensure that the time-constant of the filter is always less than the period of the highest Fourier component in your signal. If this is not so the signal will be distorted.

In the above example we have assumed that the only noise is 'white noise'. However we know that, particularly at the input of a DC amplifying system, *drift* is a problem Remember that drift is one example of $1/f$-noise. To see how we can cope with this, work through the next exercise.

∏   A system consisting of a photo-cell and amplifier is used to measure the intensity of light. The noise level in the system is shown at different frequencies in Fig. 3.5e. You can see there is $1/f$-noise together with white noise.

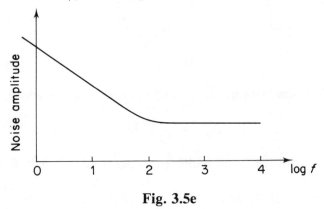

Fig. 3.5e

Explain how the effect of $1/f$-noise could be reduced if,

(i)   the light-path is chopped at a frequency, f_m,

(ii) the output from the electronic signal is passed through a high-pass filter with $f_l < f_m$.

Suggest a value for f_m.

If we do not modulate the light signal then we are making a DC measurement. This is subject to the high noise levels from $1/f$-noise which occur near to zero frequency. If the signal is chopped at a frequency, f_m, the fundamental frequency of the Fourier Transform of the signal moves from DC to a frequency, f_m. If the filter removes the low-frequency noise (less than f_l) but allows the signal to pass, it is possible to eliminate most of the $1/f$ noise.

We would need to make sure that the frequency, f_m, was greater than the frequency at which $1/f$ noise is no longer dominant. This occurs at about 100 Hz. Ideally we would like $f_m > 100$ Hz.

So far we have seen how we can reduce noise either by

— selectively eliminating certain frequencies by filters, or

— by 'moving' a signal to a quieter frequency by using modulation techniques.

The third major category of methods used to reduce noise relies on the fact that noise is random but that an analytical signal has a regular pattern of some form, see Section 3.5.2. The 'regularity' of signal may be detected by many different methods. However this category of noise reduction is beyond the scope of this Unit.

SAQ 3.5a	Explain the fundamental difference between most noise signals and analytical signals, apart from the simple fact that we do not want the noise but we do want the analytical signal.

SAQ 3.5a

SAQ 3.5b   Decide which of the following statements are true and which are false.

(*i*)   Johnson noise in a circuit can be eliminated by cooling the circuit to 0 °C.

(*ii*) White noise occurs only at optical frequencies.

(*iii*) White noise has equal components at all frequencies.

(*iv*) White noise can be eliminated by using appropriate filters.

(*v*)   Modulation of a DC signal can be used to reduce $1/f$-noise.

SAQ 3.5b

Summary

The concept of the 'Fourier Transform' is introduced in a simple non-mathematical treatment. By using this concept, the characteristics of analytical signals and noise are discussed in terms of their frequency components. This then leads logically to a discussion of the performance of various types of electronic signal processing circuits in instruments (eg amplifiers and filters).

We are then also able to discuss more advanced types of signal processing which involve 'modulation' of the initial analytical signal.

The last section examines the problems of noise more closely, and, by considering signal-to-noise ratios, consolidates the material from the earlier sections.

Objectives

It is expected that, on completion of this Part, the student will be able to:

- use the concept of a Fourier transform in analysing simple signals,

- use the frequency response characteristics to describe the performance of electronic and instrumental systems,

- identify the types of amplifier required for particular signal processing functions,

- identify the types of filter required for particular signal processing functions,

- explain the advantages of 'modulating' a signal,

- describe some techniques of modulation,

- describe the technique and requirements of a 'phase sensitive detection' system,

- explain the difference between 'signals' and 'noise',

- describe the characteristics of the principal types of noise.

4. Data-handling Elements

Overview

This part deals with the performance of some of the functional elements in instruments which affect the interpretation, presentation or passage of data. Specific examples are chosen to illustrate the type of limitations that may be imposed by the instrument on the handling of data and the extent to which this is under the control of the operator or the instrument designer.

4.1. INTEGRATION AND DIFFERENTIATION

4.1.1. Introduction

We introduced the Fourier Transform of a signal in Section 3.1.4, and investigated how electronic circuits can treat different parts of the transform in different ways. We saw in Section 3.3.3, how a low-pass filter could block high frequency components in the noise spectrum but still allow the low frequency components in the analytical signal to pass, thus improving the signal-to-noise ratio.

There are other signal-handling techniques which treat the different components of the Fourier spectrum in such a way as to enhance the desired analytical signal. In particular there are two techniques that have been in use for a number of years and are of fundamental importance – *integration* and *differentiation*. Before we discuss these two techniques individually we shall first investigate the general effect of integration and differentiation on the Fourier Transform of a signal.

4.1.2.   Effect on Fourier Components

One of the strengths of Fourier-Transform theory is that it has shown us that it is possible to look at the effect of an electronic circuit on a complicated signal by considering the *combined* effort of that circuit on the *individual* frequency components. We therefore start our investigation of differentiation and integration by looking at a single component of frequency, f.

For convenience we shall use the angular frequency, ω, given by, Eq. (4.1).

$$\omega = 2\pi f \tag{4.1}$$

We shall assume that the amplitude of this component is given by Eq. (4.2).

$$V_{in} = A \sin(\omega t) \tag{4.2}$$

The integral of V_{in} with respect to time is given by Eq. (4.3).

$$\int V_{in}\, dt = A \int \sin \omega t = \frac{-A}{\omega} \cos \omega t + \text{constant} = V_0 \tag{4.3}$$

Differentiating V_{in} with respect to time gives us Eq. (4.4).

$$dV_{in}/dt = A\omega \cos(\omega t) = V_0 \tag{4.4}$$

Apart from a change of phase, which has turned sine into cosine, and a sign change for integration, we have the following expressions for the gain function, $G = V_0/V_{in}$.

$$G = V_o/V_{in} \propto -1/\omega \qquad \text{*for integration*}$$
$$V_o/V_{in} \propto \omega \qquad \text{*for differentiation*}$$

(4.5)

These gain functions are plotted as functions of frequency in Fig. 4.1a(*i*) and (*ii*).

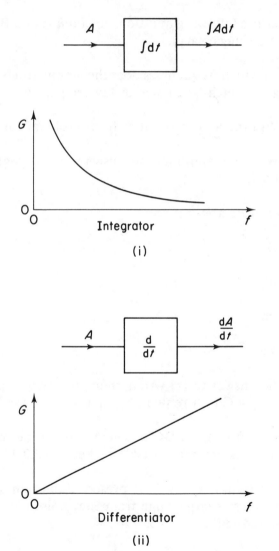

(i)

(ii)

Fig. 4.1a. *Frequency response for integration and differentiation*

The following exercises will help to bring out the significance of these two curves.

∏   There are two amplifiers, A and B. Amplifier A has the gain characteristics as in Fig. 4.1a(*i*) and amplifier B has the characteristics as in Fig. 4.1a(*ii*).

Answer the following questions for *both* A and B (refer back to Part 3 if necessary).

(*i*)   At which frequencies does the amplifier have maximum gain – high frequency or low frequency?

(*ii*)   Will the amplifier suffer from problems of drift?

(*iii*) Will the amplifier be susceptible to high-frequency noise?

The correct answers are:

	A	B
(*i*)	low	high
(*ii*)	yes	no
(*iii*)	no	yes

If you have the answer to (*i*) wrong then you have some problems in understanding a Gain/Frequency graph, see Section 3.2.3.

We know that drift (*ii*) is a DC problem and will be present if the gain of the circuit is not *zero* at DC, see Section 3.2.4.

Noise at certain frequencies will be present if the gain of the circuit at those frequencies is large. High frequency noise (*iii*) occurs only with amplifier B, see Section 3.5.4.

We shall, in fact, see below how we can use an integrator to reduce the amount of noise in a DC signal. We shall also see that a differentiator can be used to enhance the high frequency components of an analytical signal, thereby accentuating narrow lines as compared with broad lines.

4.1.3. Integrators

If a voltage V_{in} is applied to the input of an *integrating* circuit, then the output, V_o, from the circuit after a time, T, is given by Eq. 4.6.

$$V_o = (1/\tau) \times \int_0^T V_{in}\, dt + V_o(0) \qquad (4.6)$$

T is the Integration Time. $(1/\tau)$ is a constant factor for the circuit which may be called the Integrator Constant. $V_o(0)$ is the value of V_o at the start of the integrating period.

We can use the integrator circuit in two principal ways;

(*a*) If the input voltage, V_{in} is a constant DC signal then the integrator can give an improvement in the S/N ratio.

(*b*) Alternatively an integration of a spectral line as in Fig. 3.1j will give the area under the line. You probably already know from mathematics that integration gives the area under a curve. The same applies when a signal, in the form of a spectral line is fed into an integrator. The output, V_o, is proportional to the area under the peak for the spectral line. In certain situations this area is directly related to the concentration of the sample.

If V_{in} is a constant DC signal, V_{dc}, then the output rises at a constant rate such that, starting from 0 V [$V_o(0) = 0$], the output voltage would be given by Eq. (4.7).

$$V_o = (1/\tau) \times V_{dc} \times T \qquad (4.7)$$

As T increases so V_o, also continues to increase. The *rate* at which V_o increases depends on the Integrator Constant $(1/\tau)$. This may be a controllable parameter similar to 'gain' in a normal amplifier. The effect is that a constant DC signal is added up over a period of time, T. During that same time the noise signals which occur randomly with positive and negative values will tend to average out. The combination of these two factors will give an improvement in the S/N ratio. If we increase the length of the integration time we can continue to increase the S/N ratio.

For the noise signal at a particular frequency, f_n, to average-out, that noise signal must pass through at least one complete cycle. This means that the integration circuit will reduce noise components only with frequencies, f_n, such that

$$f_n > 1/T$$

where T is the integration time involved. Thus the integrating circuit tends to reduce all noise except for very-low-frequency noise. The fact that low-frequency noise is not eliminated means that an integrator is *very* susceptible to *drift*. It will continue to integrate small drift-voltages as though they were actual signals. These characteristics of an integrating circuit can be understood by looking back to its Fourier Transform in Fig. 4.1a(i).

In the use of an integrator, an integration time must be chosen so that it is as long as is conveniently possible without developing severe drift. The signal input must be kept sufficiently low that, over the integration time, the required output V_o does not exceed its possible voltage range. If the signal magnitude is too large, it may be passed through an attenuator which will reduce it to an acceptable value.

∏   In a particular Atomic Absorption spectrophotometer, an integrator system is used at the output of the instrument to enhance the S/N ratio. This system essentially consists of an attenuator with possible settings of × 1, × 0.1, and × 0.01, and an integrating circuit with a fixed integrator constant.

If the attenuator is set at × 0.1, and with an integration time of 0.10 s, an input voltage of 90 mV gives an output voltage of 9.0 V. By assuming also that the maximum (saturation) value for the output of the integrator is 10 V, calculate the output voltages for the following conditions.

	(i)	(ii)	(iii)	(iv)
Input voltage/mV	90	9	0.9	1.1	1.1
Attenuation	× 0.1	× 0.1	× 0.1	× 0.1	× 0.01
Integration time/s	0.1	1.0	10	10	10
Output voltage/V	9	..	..	..	..

The output, V_o, from the complete integrator system can be written as follows

$$V_o = G \times (1/\tau) \times V_{dc} \times T$$

where V_{dc} is the constant input and T is the integration time, $(1/\tau)$ is the integrator constant of the integrator. G is the 'gain' of the attenuator.

Using the initial values given,

$$9.0 = 0.1 \times (1/\tau) \times 0.09 \times 0.1$$

This gives $(1/\tau) = 10000 \text{ s}^{-1}$.

Using this value of $(1/\tau)$ for the different settings gives, for

(i) 9.0 V,

and for (ii) 9.0 V.

Hence a reduction in the input voltage (ie a weaker signal) can be compensated by using a longer integration time.

If the above equation is used without thinking for (iii), ie for a larger signal time than (ii), then the answer would appear to be 11 V. However, this is not possible because the output from the system is limited to 10 V. This situation is called 'saturation' or 'overload', and the reading would remain at 10 V irrespective of its true value. Therefore a greater attenuation must be used, as in (*iv*) to bring the result into an acceptable range – 1.1 V.

In practice when you increase the integration time in an instrument then the attenuation factor is often automatically increased to *prevent saturation.*

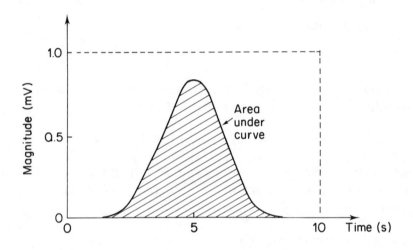

Fig. 4.1b

∏   Fig 4.1b shows the variation with time of the voltage level at the output of a Gas Chromatograph due to a particular output signal. The area under the curve for the signal is equal to 2.5×10^{-3} V s. This signal is fed into the same integrator system as in the previous exercise with an attenuator setting of $\times\ 0.1$ and an integrating time of 10 s.

Calculate the voltage output from the integrator after the signal has passed.

Explain why $\times\ 0.1$ is now the optimum setting for the attenuator.

We know from the previous exercise that the output, V_o, for a DC input is given by the equation below,

$$V_o = G \times (1/\tau) \times V_{dc} \times T$$

where $G = 0.1$, and $(1/\tau) = 10000 \text{ s}^{-1}$.

If however, V_{dc} is a voltage, V, which changes with time then,

$$V_o = G/\tau \times \int_0^T V dt$$

The integral is equal to the area under the curve, with voltage and time as the two axes.

We are told that the area under the curve signal is 2.5×10^{-3} V s.

$$\int_0^T V dt = 2.5 \times 10^{-3} \text{ V s}$$

$$\therefore \quad V_o = 0.1 \times 10000 \times 2.5 \times 10^{-3} = 2.5 \text{ V s}$$

If the attenuator had been set to $\times 1$, then the output would have been saturated (at 10 V). If the setting had been at $\times 0.01$, then the output voltage would have been unnecessarily low at 0.25 V.

4.1.4. Differentiator

If a voltage V_{in} is applied to the input of a *differentiator* circuit, then the output, V_o, will be given by Eq. (4.8).

$$V_o = \tau \times d V_{in}/dt \tag{4.8}$$

The constant tau is a Time Constant for the particular circuit. V_o will depend on the rate at which V_{in} is changing. Hence the differentiator favours the high frequencies over the low frequencies. Obviously if V_{in} is a constant voltage, V_{dc}, then the output will be zero.

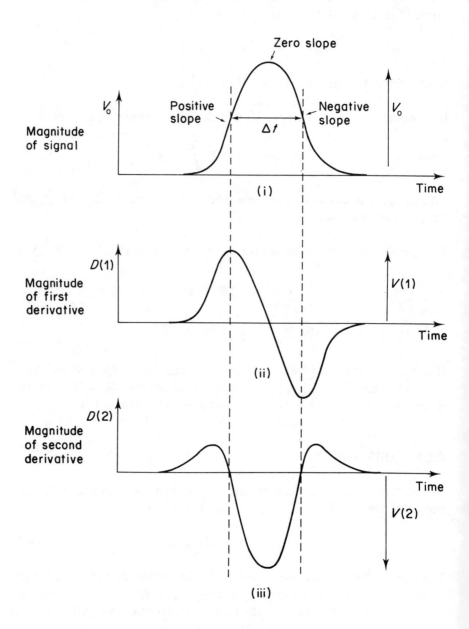

Fig. 4.1c. *Differentiation of line signal*

If we consider a 'spectral line' as in Fig. 4.1c(i), which may have been drawn out on a chart recorder as a graph of V_0 *versus* t, the voltage, V, is then a function of time, t.

We can differentiate this function to obtain,

$$D(1)V = dV/dt$$

$D(1)V$, the first derivative of V, is given by the *slope* of the graph of V against t, so that we can obtain the graph of $D(1)V$ just by measuring the slope of V at every point in time, see Fig. 4.1c(ii).

This differentiating process can be repeated to obtain the second derivative, $D(2)V$, where

$$D(2)V = d(D(1)V)/dt = d^2V/dt^2$$

This is given in Fig. 4.1c(iii).

The advantage in doing this is that it accentuates the presence of small narrow spectral lines on the shoulders of broad lines. The following exercises illustrate this point.

∏   If the spectral line given in Fig. 4.1c has an almost Gaussian shape then it is possible to give approximate relationships between the amplitudes, V_0, V_1 and V_2. The width at half-height of the signal is Δt s.

In each of parts A and B below, there is only one equation which is approximately correct.

Indicate which are true and which are false.

A

(i)	$V(1) \propto V_0 \times \Delta t$	T / F
(ii)	$V(1) \propto V_0/\Delta t$	T / F
(iii)	$V(1) \propto \Delta t/V_0$	T / F

B

(i) $V(2) \propto V_0^2 \times \Delta t$ T / F

(ii) $V(2) \propto V_0^2/\Delta t$ T / F

(iii) $V(2) \propto (\Delta t)^2/V_0$ T / F

(iv) $V(2) \propto V_0/(\Delta t)^2$ T / F

The correct answer is

A		B	
(i)	F	(i)	F
(ii)	T	(ii)	F
(iii)	F	(iii)	F
		(iv)	T

Although we have not given a detailed mathematical treatment, it is possible to understand the basic relationship as below.

If V_0 is zero then $V(1)$ and $V(2)$ must also be zero. Thus an increase in the size of the original signal, V_0, must result in an increase of both $V(1)$ and $V(2)$. Hence A (iii) and B (iii) must both be false.

We know that *reducing* the width (Δt) of the line whilst maintaining the same amplitude will give a greater slope and hence *increased* values for $V(1)$ and $V(2)$. This is achieved only by making Δt the divisor in the equation. Thus A (i) and B (i) are also false.

Hence A (ii), being the only remaining equation in A must be the correct one.

If we apply the equation,

$$V(1) \propto V_0/\Delta t$$

putting $V(2)$ as the differential coefficient of $V(1)$, then we obtain combining these equations

$$V(2) \propto (V_0/\Delta t)/\Delta t = V_0/(\Delta t)^2$$

Hence B (*iv*) is true
and   B (*ii*) is false.

Note that the magnitude of D(1) and of D(2) in Fig. 4.1c depends on the *rate with respect to time* at which the spectral line is drawn out.

If the spectral line is produced from a scanning spectrophotometer then we would normally expect to see it drawn as a function of wavelength as in Fig. 4.1d. The rate at which the line is drawn out in time then depends on the speed at which the spectrophotometer is scanning the wavelengths.

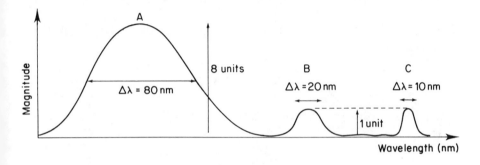

Fig. 4.1d

Π   Referring to the spectral lines A, B and C in Fig. 4.1d, the second derivative signal for line A is found to have a magnitude, $V(2)$, of 4.0 units when the spectrum is scanned at a speed of 10 nm s^{-1}.

Calculate the magnitude of the second derivatives, $V(2)$, for lines A, B and C for the different scan-speeds given below.

Scan-speed (nm s^{-1})	$V(2)(A)$	$V(2)(B)$	$V(2)(C)$
10	4.0	...	...
20	...	...	...

The important equation in calculating the results is obtained from the previous exercise.

$$V(2) = K \times V_0/(\Delta t)^2$$

where K is a constant.

V_0 for A, B and C is 8, 1.0 and 1.0 V respectively.

To obtain the values of Δt we need first to take the values of line-width from the spectrum.

$\Delta \lambda$ for A, B and C is 80, 20 and 10 nm respectively.

The values of Δt are obtained by using the scan speed, S, and the relationship:

$$\Delta t = \Delta \lambda / S$$

This gives values for Δt and $(\Delta t)^2$ as below,

Scan speed nm s^{-1}	Δt/s			$(\Delta t)^2$/s^2		
	A	B	C	A	B	C
10	8	2	1	64	4	1
20	4	1	0.5	16	1	0.25

By using the initial conditions for line A and substituting into the equation above we obtain:

$$4.0 = K \times 8/64$$

$$K = 32$$

Substituting this for lines A, B and C at both scan speeds we get the values given in the answer table below.

Scan-speed (nm s^{-1})	V(2)(A)	V(2)(B)	V(2)(C)
10	4.0	8.0	32.0
20	16.0	32.0	128.0

Note that by using the *second* derivative, the signal is enhanced as the *square* of the reduction in the time taken to pass through the line. Thus C, which is only one eighth of the amplitude of A can in second derivative, produce a signal eight times larger than A, because of their difference in line-widths.

We have seen in these exercises that a differentiating circuit will amplify signals by a factor which is inversely proportional to their period. High-frequency noise has a very short period, and this will *also* be amplified by the differentiating circuit. Unless we take care we would find that the analytical signal would be swamped by lots of noise of very high frequency. The solution is to include a low-pass filter with a high-frequency cut-off at a frequency just greater than the highest component in the analytical signal. In this way excessive high-frequency noise can be minimised. This low-pass filter is often built into the instrument,and may even be due to the time-constant of the instrument itself. However, it requires some skill to obtain the optimum setting, and you should always check to ensure that your results are not unduly affected by instrumental noise.

4.1.5. Conclusion

We have seen in this section that we can emphasise different aspects of a complex analytical signal by modifying the Fourier Transform of the signal. We have also seen that such a system is also liable to emphasise different types of noise.

An integrator emphasises DC and low-frequency components of a signal and is therefore also susceptible to drift.

A differentiator emphasises high-frequency components and is susceptible to these frequencies in the noise spectrum.

Modern computing methods now allow for many techniques in data handling. Such developments are beyond the scope of this unit, but, after studying the examples given in this text, you should now be more aware that, although such techniques are able to bring out valuable information from a signal, they also require understanding and careful operation to prevent emphasising instrumental errors.

SAQ 4.1a

Complete the following sentences by using phrases from the list given below.

(*i*) .... can be a serious problem in an integrating system.

(*ii*) The output from a differentiator circuit can easily suffer from ....

Possible phrases:

(*a*) wavelength scan,
(*b*) attenuation,
(*c*) high-frequency noise,
(*d*) AC drift,
(*e*) amplification,
(*f*) DC drift.

SAQ 4.1b

With reference to a wavelength spectrum which contains a broad line, A, and a narrow line, B, $V(A)/V(B)$ is the ratio of the magnitudes of the two lines in the normal spectrum, and $V(2)(A)/V(2)(B)$ is the ratio of the magnitudes of their second derivatives with respect to time.

Which of the following statements are true?

(*i*) $V(2)(A)/V(2)(B) < V(A)/V(B)$ T / F
(*ii*) $V(2)(A)/V(2)(B) > V(A)/V(B)$ T / F
(*iii*) $V(2)(A)/V(2)(B) = V(A)/V(B)$ T / F
(*iv*) $V(2)(A)/V(2)(B)$ will increase if the wavelengths are scanned more quickly.
(*v*) $V(2)(A)/V(2)(B)$ will decrease if the wavelengths are scanned more quickly.

4.2. PULSES

4.2.1. Introduction

An electrical pulse occurs when the voltage changes to a different value for a short time. A single pulse is an isolated change which does not recur, hence it is not possible to say that it is either a DC (constant) or AC (regular repetitive) signal.

We can use a *series of pulses* to carry information. This information can be conveyed by the pulses *via* their

amplitude,

pulse width,

or *frequency of occurrence*.

The information can also be carried as a *coded* sequence of pulses of the same amplitude. This *digital* signal was mentioned briefly in Section 2.2.8 and will be covered more extensively in Section 4.3.2. In this section we shall discuss factors affecting only *pulse height* and *frequency*.

With the advance in modern electronics, pulse systems are now very common. We saw in Section 3.4 how the signal from a modulated optical system could be converted with the aid of tuned filters into AC Fourier components. The same signal and information can now be handled more efficiently with electronic pulse-circuits. We shall, however, concentrate on physical systems that directly dictate the use of pulse electronics – the detection of high-energy radiation and particles.

4.2.2. Measurement of Streams of Photons and Radioactive Particles

Electromagnetic radiation is not a continuous stream of energy but a stream of individual packets of energy called *photons*. Quantum theory gives the energy, E, for each 'packet' as depending on the frequency, ν, of the electromagnetic wave,

$$E \;=\; h\nu \tag{4.9}$$

where h is Planck's Constant. We use ν for the electromagnetic frequency to avoid confusion with the frequency, f, at which photons arrive.

For ir, visible, and uv radiation, the size of the photon is so small that in most instruments the light beam contains so many photons that it appears to be a continuous stream of energy. These instruments record a signal as though the radiation were in fact continuous.

At higher em frequencies, eg for X-rays and γ-rays, this is no longer true. The photon-size is now so great that the arrival of each photon is recorded as separate events. Note that, for γ-rays, the familiar 'click-click' of a Geiger counter is recording separate γ-ray photons. Each high-energy photon will produce a pulse in the electronic system. The frequency, f, at which these pulses occur will depend on the rate of arrival of photons.

A detecting system can be produced such that the amplitude, V_p, of a particular pulse is a function, F, of the energy, E, of the photon that created the pulse, ie,

$$V_p \;=\; F(E) \tag{4.10}$$

This can also apply to the energy, E, of high-energy particles such as γ-particles and β-particles. We discuss below the ways in which the energy, E, and the frequency, f, can be derived from the pulsed signals.

4.2.3. Use of Pulse Height

We assume as above that the pulse height, V_p, is a simple function of the energy, E, of the photon or particle.

Π   Look at the signal in Fig. 4.2a, and try to decide how many
       pulses are visible.

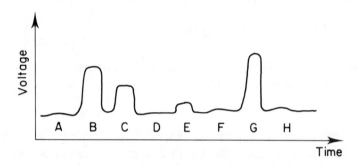

Fig. 4.2a.   *Electronic pulses*

There is no completely correct answer to this question because we
have not yet defined a minimum voltage which enables us to distinguish a pulse from noise.

There are definite pulses at B, C, E and G. We might even be con-
fident that there were small pulses at A and F, but what about H?
Is H a real pulse or is it noise? Could A and F also be noise?

The object of this exercise was to highlight the problems involved
in distinguishing between noise and a signal.

The electronic circuits must possess criteria by which they can 'decide' whether or not to accept or reject the existence of a pulse.
The counting circuit must have a minimum *threshold voltage* below which any voltage fluctuations are considered to be noise
and are rejected, ie only pulses above this threshold voltage are
counted.

We can have a more complex circuit which will respond to a pulse
only if it has an amplitude below any chosen voltage, V_h, and above
a lower voltage, V_l. This would be a *discriminating* circuit responding only to pulses whose amplitude is between V_l and V_h – a *voltage discriminator*. A counter attached to this discriminator would
then count how many pulses had voltages in a particular range. If

a sequence of pulses is fed into a number of *discriminators* each set to cover a different section of the complete voltage range of the pulses, the output from each discriminator represents the number of pulses in a certain voltage range. These different 'counts' then give a 'profile' (or spectrum) of pulse number as a function of pulse voltage.

∏   Each of the vertical lines in Fig. 4.2b represents a single pulse. Five separate discriminators each have a range ($V_h - V_l$) of 1.00 V, and cover separate sections of the pulse voltage from 1.00 V to 6.00 V.

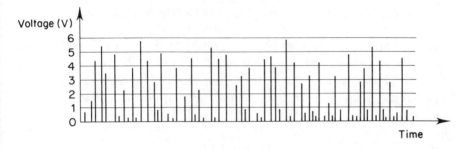

Fig. 4.2b

Count the number of pulses with amplitudes in each of the 5 ranges and draw a histogram of pulse-count against voltage.

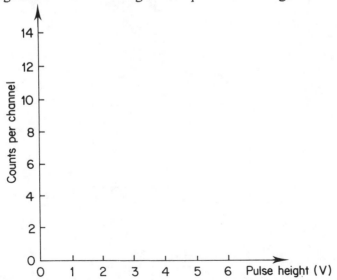

We are not counting any pulses with an amplitude below 1.00 V, so all these can be disregarded whether they are noise or even small pulses.

By counting between the limits in each range we get:

A   1–2 V range –   3 pulses,
B   2–3 V range –   7 pulses,
C   3–4 V range –   9 pulses
D   4–5 V range – 14 pulses,
E   5–6 V range –   5 pulses.

By plotting these on a histogram we obtain the diagram in Fig. 4.2c.

If you obtained a series of counts similar to 3, 10, 14, 33, 38, then you have incorrectly taken the number of pulses below each given voltage but without a lower limit. We ignore, for each range, any pulse whose amplitude is outside *both* the upper and lower limit.

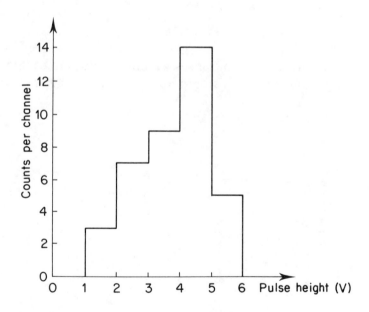

Fig. 4.2c

If we know that the height of the pulse, V_p, is a simple function of the energy of the photons, E, then the histogram is also a plot of the *number of photons* in the radiation as a function of the *energy of the photons*. The graph is then a type of *spectrum* for high-energy photons or particles.

A *Multichannel Analyser* (MCA) is an electronic data-gathering instrument designed specifically to deal with the type of situation outlined above. Instead of just using 5 recording channels as in our exercise, an MCA can typically have 1024 *channels* each counting the number of pulses arriving with a voltage within very small, but different, voltage ranges. The histogram, or spectrum can be displayed directly on a visible screen as the counts are progressing.

∏   The radiation from a particular radioactive source produces a MCA spectrum as shown in Fig. 4.2d. We know that the pulse voltage V_p is related to the energy, E, of the radiation particles, but it is also a function of the detector, the electronics, and the gain settings in the instrument. In view of the complexity of trying to equate V_p *directly* with E, it is not easy to calculate the particle energy corresponding to a given voltage. How then can we calibrate the instrument? Can you suggest a method for finding the energies corresponding to particular voltages in the spectrum?

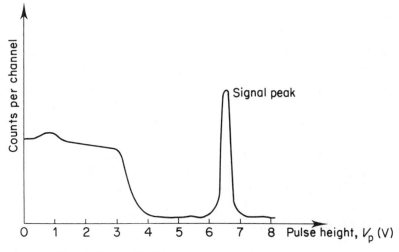

Fig. 4.2d.   *Multi-channel analyser spectrum*

This problem is one which recurs in many different forms through-out instrumental analysis. It is often far too difficult or inaccurate to try to *calculate* the action of each part of the instrumental process. Instead we can measure its effect by feeding in a spectrum from a sample whose characteristics are well known from other measure-ments. Here we would take a *known* radioactive source and observe where on our voltage scale particular emissions are recorded. Then, knowing the energies, E, of these emissions, we can use our *standard sample* to *calibrate* the scale of the instrument.

Obviously if we increase the number of channels available we shall be able to distinguish between particles or photons with small energy differences and the *resolution* in the spectrum will become better. Note that it is also necessary that the detector itself is capable of re-solving the small energy differences. However, there is one problem in improving resolution. As we put pulses into a greater number of channels, then the number of pulses per channel will correspond-ingly decrease. We must then take a longer time to record enough pulses per channel or else there will be a reduction in the accuracy of the count in each channel.

We shall deal with the problem of *counting* pulses in the next sec-tion.

4.2.4. Counting of Pulses

Counting pulses does not present any real problems in the electron-ics once the pulses are produced by the detector. The main difficulty arises in trying to relate the number of electronic pulses produced by the detector to the actual number of photons or particles arriving at the detector.

After a photon arrives at the detector and an electronic pulse is emitted, the detector must then re-establish its equilibrium before it can be ready for another photon. It will take a certain time before the detector has recovered and is ready to repeat the process. If another photon arrives during that time it will just be ignored and lost from the count. This means that for a high photon-arrival-rate, a number of photons will be missed by the detector because it has not yet recovered from a previous pulse.

∏   Fig. 4.2e shows a series of pulses coming from a detection system. After the start of each pulse, the whole system is 'dead' for a time T_d. Calculate the average rate at which photons are arriving at the detector. Note that any pulse which occurs during a time T_d after another pulse will not be counted.

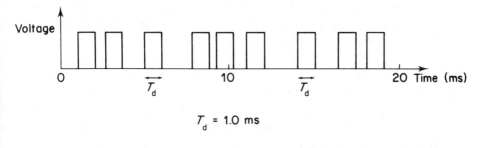

$$T_d = 1.0 \text{ ms}$$

Fig. 4.2c

Hint: Instead of using the total time elapsed, calculate the actual time for which the detector was active and waiting for pulses.

We can see from the diagram that there are 9 pulses arriving within a period of 20 ms. However, during that period the detector is 'dead' for a time $9 \times T_d = 9.0$ ms. The detector is available to respond to pulses only within a period of $20.0 - 9.0 = 11.0$ ms.

We then know that 9 pulses arrive within a period of 11.0 ms, giving a rate of 0.82 pulses ms.

In fact the time for which a detector is 'dead' may vary from pulse to pulse, and this variation would make it difficult to estimate how many pulses were missed. This can be overcome by artificially imposing a constant system-'dead time', T_d, which is always longer than the actual 'dead-time' of the detector, but since it is a constant time, its effect can be included in any calculation. By using a constant dead-time, T_d s, it is possible to estimate statistically the actual photon arrival rate, N_p (in counts s^{-1}) based on any particular count rate, N_c, by using Eq. 4.11.

$$N_p = N_c/(1 - T_d N_c) \qquad (4.11)$$

4.2.5. Statistics of Counting

The arrival of either photons or radioactive particles is not regular. Each arrival is a statistically random event. If a fixed time is taken for counting these events then the actual number counted also has a statistically variable value.

If, over a certain length of time, the *average* count is N_0, then the standard deviation, s, of the possible counts would be $\sqrt{N_0}$. This can be illustrated by using an example where the average count, N_0, is 14400. Fig. 4.2f gives the relative probability with which we expect actually to record a count of N. This is called a Normal Distribution.

In our particular example, the standard deviation, s is given by:

$$s = \sqrt{N_0} = \sqrt{14400} = 120.$$

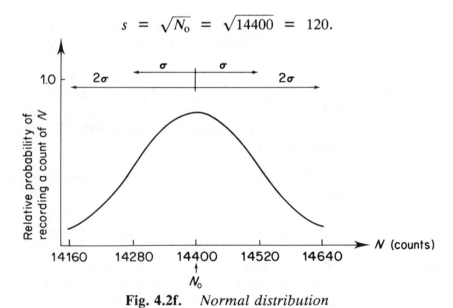

Fig. 4.2f.   *Normal distribution*

In a Normal Distribution we know that any particular count has a 68% probability of being within one standard deviation of the average value of many such counts. Thus the standard deviation is a good measure of the *statistical error* involved in a particular measurement. We do not propose to go deeply into statistics here. This will be dealt with in greater depth in units concerned with the treatment of results and errors.

∏   By using one standard deviation as the measure of error, calculate the *percentage* error in a count of 100.

  If the counting period is increased fourfold, estimate the new *percentage* error in the count.

For an actual count of $N = 100$ we know that N_o will also be close to 100. Thus we can expect that the standard deviation will be $\sqrt{N_o}$ = 10.

Thus if our actual count is 100 we can expect that the average value of repeating the count many times, N_o, would be 100 ± 10, with a probability of 68%.

  The statistical error is ± 10,

  percentage error is $\pm 10/100 \times 100 = \pm 10\%$

If the count period is increased fourfold then we would expect a count of about 400.

Using the same method as above this gives us:

  Statistical error is $\pm \sqrt{400} = \pm 20$,

  Percentage error is $\pm 20/400 \times 100 = \pm 5\%$

Thus increasing the count period by a factor of x will reduce the *percentage* error by a factor equal to the square-root of x.

In terms of the concept of signal-to-noise ratio, S/N, introduced earlier, the 'signal' is the average count rate, N_0, and the 'noise' is the statistical error, $\sqrt{N_0}$.

Performing a count over a period of time is equivalent to an *integration* (see Section 4.1) of the count rate. We can see that increasing the period of integration improves the signal-to-noise ratio of the result.

You will have noticed that the S/N ratio in counting improves as the square-root of the increase of the integration period. You may also remember that, for a signal mixed with 'white' noise, the S/N ratio can be improved as the square-root of the time-constant of a low-pass filter. The similarity between these two situations is not surprising because for both, the noise is due to statistically random events. We use randomness of the noise as a means by which we can reject it in comparison to the desired signal.

SAQ 4.2a	Assume that, when we are recording the energy spectrum of a beam of γ-rays with a multichannel analyser, we decide to use 1024 channels instead of 512 for the same range of γ-ray energies and for the same counting period.

Which of the following statements are true?

Increasing the number of channels means that:

(*i*) it is easier to distinguish between photons of nearly the same energy,

T / F

(*ii*) the signal-to-noise ratio of the spectrum is increased,

T / F

⟶

**SAQ 4.2a
(cont.)**

(*iii*) the signal-to-noise ratio of the spectrum is reduced,

T / F

(*iv*) we can halve the time taken to record the spectrum without reducing the signal-to-noise ratio.

T / F

SAQ 4.2b

Detection of a particular beam of X-rays gives a count rate of 40 counts s^{-1}.

Calculate the percentage error (by using one standard deviation) in the result for a counting period of 10 s.

What counting period would be required to reduce the percentage error to one half of this value? (You should be able to do the second part of this question very quickly without very much calculation.)

SAQ 4.2b

4.3. INTERCONVERSION OF DIGITAL AND ANALOGUE SIGNALS

4.3.1. Introduction

The use of digital systems and microcomputers is of such importance in modern chemical instrumentation that a complete unit is devoted to the topic. However, it is important for those students who might not yet have studied that unit to be aware of some of the main problems that may arise when we use digital signals.

4.3.2. Digital Representation of an Analogue Voltage

The idea of a digital signal was introduced in Section 2.2.8 when we were dealing with *electrical information*. In that section we saw that the exact amplitude of the digital pulse did not carry any information. The information in each pulse was limited to the existence or otherwise of the pulse – either '0' or '1' in terms of a binary code. A *digital* signal conveys *numbers* expressed in terms of a *binary* code.

We can represent an analogue voltage (say 3.0 V) by a digital signal provided that we are able to write the *decimal* number (3.0) as a *binary* number.

A binary 'number' is a sequence of 'bits', each 'bit' representing an increasing power of *two*. For an 8-bit binary number this gives us a series of bits which we can identify below.

Bit	Bit magnitude
$b0$	$2^0 = 1$ Least significant bit
$b1$	$2^1 = 2$
$b2$	$2^2 = 4$
$b3$	$2^3 = 8$
$b4$	$2^4 = 16$
$b5$	$2^5 = 32$
$b6$	$2^6 = 64$
$b7$	$2^7 = 128$ Most significant bit

Thus, for example, the number '3' in decimals would, in an 8-bit binary system, become 0000 0011, and decimal 4 would become 00000100.

A binary number containing 8 bits is called a *byte*. The bit, $b0$, is called the '*Least Significant Bit*' or LSB because it represents the smallest unit in the byte. The bit $b7$, would be called the '*Most Significant Bit*', or MSB. The largest 'number' that can be expressed in an 8-bit binary system would be 1111 1111 which is equivalent to the decimal number 255.

∏ (*i*) Express the binary number 1001 1001 (with the LSB on the right) as a decimal number.

   (*ii*) Express the decimal number 102 as an 8-bit binary number.

(*i*) The 'code' in the binary number, 1001 1001, has $b7$, $b4$, $b3$, $b0$ all equal to 1 and $b6$, $b5$, $b2$, $b1$ all equal to 0.

This tells us that the decimal number is made up of:

$$1 \times 128 + 1 \times 16 + 1 \times 8 + 1 \times 1 = 153$$

Binary 1001 1001 is equivalent to decimal 153.

If you don't know where the numbers 128, 16, 8 and 1 come from for $b7$, $b4$, $b3$ and $b0$ respectively then check back to the text.

In general the decimal number is given by

$$b7 \times 128 + b6 \times 64 + \ldots + b0 \times 1$$

where the values of $b7$ to $b0$ are given by the binary coded number.

Clearly the conversion from binary into decimal is very straightforward, however the reverse process involves a more complicated procedure as we see below.

(*ii*) There are several ways that the conversion from decimal into binary can be done, and here we choose to use a method of successive approximation.

We wish to find a binary number $b7$, $b6$, $b5$, $b4$, $b3$, $b2$, $b1$, $b0$ that is equivalent to decimal 102.

The first question is whether $b7$ is '0' or '1'. Since $b7$ is equivalent to 128 on its own and this is more than our decimal number (102), then $b7$ can not be allowed to have the value '1'. It must be '0'.

However, $b6$ is equivalent to 64, so if we set $b6 = $ '1', this will leave $102 - 64 = 38$ to be distributed amongst the other 6 bits ($b5$ to $b0$). Similarly $b5$ will be '1', leaving $38 - 32 = 6$. Then $b4$ ($= 16$) and $b3$ ($= 8$) must both be '0'. We can see quickly that decimal 6 is equivalent to 110 for the last three bits.

Hence,

Binary 0110 0110 is equivalent to decimal 102.

In this particular case you may have been quick to notice that 153 from (*i*) and 102 (*ii*) add up to 255, or

$$255 - 153 = 102$$

In binary form this is:

$$1111\ 1111 - 1001\ 1001 = 0110\ 0110.$$

If you did it this way, well spotted, but you should also know the more general way described above.

What does all this mean when we want to convert an analogue voltage into an 8-bit binary number so that we can send it along a wire as a digital signal? If, in a very simple case, we had an analogue signal that could have any possible value between 0 V and 255 V, then the conversion should be fairly clear. We would make the LSB equivalent to 1 V. Then,

00000001 would represent 1 V

00000010 would represent 2 V

....

11111111 would represent 255 V

The example given is, however, somewhat unrealistic in that it is unlikely that the analogue voltage would have a *range* of exactly 255 V. It is more likely that it would have a range of perhaps 1.0 V. Obviously we must then make the LSB equivalent to a voltage of < 1 V.

∏   If a digital signal of 0000 0001 (LSB on the right) is used to *represent* an analogue voltage of 4.00 mV, calculate the voltage which is represented by

(*i*) 1111 1111,

(*ii*) 1001 1001.

In this question we are concerned with two-step conversions; from binary to decimal numbers, and the from a decimal number to a voltage.

0000 0001 is equivalent to decimal 1. Hence decimal 1 is equivalent to a voltage of 4.00 mV.

(*i*)   Binary 1111 1111 is equivalent to decimal 255, which is then equivalent to 255 × 4.00 mV = 1020 mV = 1.020 V.

(*ii*) Binary 1001 1001 gives decimal 153. The equivalent analogue voltage is then obtained by multiplying by the size of the LSB.

$$= 153 \times 4.00 \text{ mV} = 0.612 \text{ V}.$$

If in the above question you did not get the *decimal* numbers (255 and 153) then check back to the previous exercise.

For a voltage range of 1.00 V a convenient choice for the size of the LSB would be 4.00 mV, then,

00000001 would represent 4.00 mV

00000010 would represent 8.00 mV

00000011 would represent 12.00 mV

....

10000000 would represent 512.00 mV

....

11111111 would represent 1020.00 mV

= 1.020 V

∏   If the LSB of an 8-bit signal is equivalent to 4.00 mV, calculate:

   (*i*)   the digital signal equivalent to 0.408 V,

   (*ii*) the (best) digital signal equivalent to 0.410 V.

This is the reverse of the process to that used in the previous exercise. We must find a decimal number *equivalent* to the voltage with 4.00 mV as being *equivalent* to decimal 1.

(*i*)   If 4.00 mV is equivalent to 1 then 0.408 V (= 408 mV) is equivalent to 408/4 = decimal 102.

We know that decimal 102 is equivalent to binary 0110 0110.

∴   0.408 V would be represented by binary 0110 0110.

(*ii*) A voltage of 0.410 V is 2.00 mV greater than the 0.408 V in (*i*). 2.00 mV is not equivalent to an exact multiple of the LSB (= 4.00 mV) so we cannot assign an exact binary equivalent.

We could choose:

       0110 0110 (= 0.408 V) – too *low*

or   0110 0111 (= 0.412 V) – too *high*.

∴   There is an error of ±0.5 LSB.

You should now be aware of one source of error in using a digital system. It is not possible, using an 8-bit number over a range of 1.02 V to express the voltage in steps smaller than 4.00 mV. If we have, for example, an analogue voltage of 18.0 mV then we must represent it, either as 00000101 which is equivalent to 20.0 mV, or as 00000100 which is the equivalent to 16.0 mV. In every digital number there is a *minimum inherent error* equal to ± 0.5 × LSB. This is called the *Quantisation Error of Conversion*.

The quantisation error for an 8-bit number with a range of 1.0 V is ±2.0 mV.

∏   An analogue voltage, V, is converted into a digital signal. The *quantisation* error of conversion is ±2.00 mV (which is equivalent to a LSB with a value of 4.0 mV).

Calculate the equivalent *percentage* error in V if the voltage V is:

(*i*)   about 1 V,

(*ii*)   about 0.1 V,

(*iii*)   about 0.01 V.

In all parts of this question the actual error magnitude is the same, ie ±2mV, but the *percentage* error depends on the relative sizes of signal voltage and error.

(*i*)   Signal = 1000 mV Error = (2/1000) × 100 = 0.2 %
(*ii*)   Signal = 100 mV   Error = (2/100) × 100   = 2 %
(*iii*) Signal = 10 mV   Error = (2/10) × 100   = 20 %

Notice that in each case the voltage of the signal has been expressed in millivolts to get an appropriate ratio with the 2.0 mV error.

Note in particular that if the voltage of the signal is small compared to the voltage *range* of the digital signal, there can be a large percentage error in conversion.

The effect of quantisation error is most serious when the signal being measured is *much smaller* than the *total possible range*.

4.3.3.   Digital-to-Analogue (DTA) Conversion

Conversion *from* a digital signal *into* an analogue signal is fairly straightforward in both the mathematical calculation (see the second exercise in this section) and in the electronic circuit (at least in principle).

If the input is an 8-bit digital signal (see Fig. 4.3a) and the output has a maximum *range* of 1.02 V, then, following the example above, the LSB of the input must be equivalent to an output of 4.0 mV.

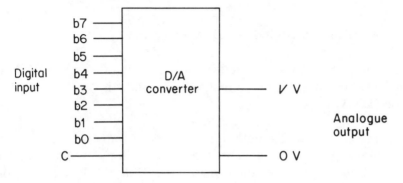

Fig. 4.3a. *Digital to analogue conversion*

A '1' on the $b0$ line must add 4.0 mV to the output.

A '1' on the $b1$ line must add 8.0 mV to the output.

A '1' on the $b2$ line must add 16.0 mV to the output.

···

A '1' on the $b7$ line must add 512.0 mV to the output.

A '0' on the input adds nothing to the output.

The final output voltage, V_o, is the sum of all of these contributions.

It may be possible to change the 'gain' of the circuit. This will change the magnitude of the output voltage which is equivalent to the LSB of the input. However, if we do increase the gain of the circuit we must remember that we are also increasing the size of the error due to the Quantisation Error of Conversion (0.5 LSB).

When a new digital signal is fed into an DTA converter, it will obviously take a certain time for the output to stabilise at the new correct value. The time it takes for the voltage to stabilise within a voltage equivalent to ±0.5 LSB is called the *Settling Time*. Usually this is very short, ie less than one microsecond.

4.3.4.   Analogue-to-Digital (ATD) Conversion

The conversion of an analogue signal into a digital signal (ATD) is more complicated, both mathematically and electronically, than the reverse (DTA) process.

We saw in Section 4.3.2 a method of *successive approximation* to obtain a binary number equivalent to a given analogue voltage. There are electronic circuits which use this method for the conversion process.

There are also other ways. One of these would be to start from a voltage equal to the LSB in the binary number, and then start writing down every ascending binary number and its equivalent voltage.

   0000 0001 = 4 mV

   0000 0010 = 8 mV etc

At each step you would check to see whether your equivalent voltage was still less than the given input analogue voltage. When the two voltages are within 0.5 LSB (within 2 mV in this example) then you will find that you have just written down the necessary binary number. Although this sounds very tedious to use it is a common method of producing one type of ATD converter.

Both the methods suggested for ATD conversion require a few calculation steps. This takes time and you should now appreciate that analogue-to-digital (ATD) conversion can be a time-consuming business. This can be anywhere between a microsecond to several milliseconds depending on cost and applications. The actual *Conversion Time* of an ATD circuit is an important part of the specification of its performance, and it is a limiting factor that you should be aware of if you are dealing with an analogue voltage which is rapidly changing from one value to another.

4.3.5. Sample-and-Hold Circuit

We now know that the actual process of ATD conversion takes a finite time – the conversion time. During that time the analogue voltage applied to the converter ideally should not change. If it does, then the ATD circuit may well become confused. However we shall frequently be dealing with voltages that do indeed change. What must be done?

The answer is to feed the varying analogue signal into a *Sample-and-Hold* circuit. This does exactly what it says. The value of the analogue signal is sampled at one instant in time and that voltage is held constant and passed into the input of an ATD converter. When conversion of that particular sample has been completed, the 'sample-and-hold' circuit can then be instructed to take a new sample of the value of the analogue signal. Any values that the analogue signal may have between sample times are completely ignored. See Fig. 4.3b.

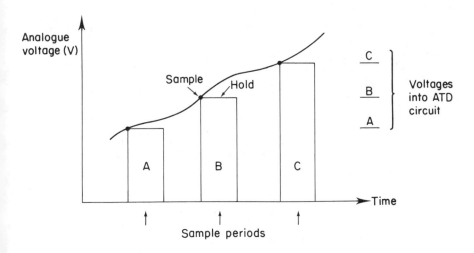

Fig. 4.3b. *Sample-and-Hold measurement*

One hopes that in any well-designed system the *sampling rate* is very great compared with the speed at which the analytical signal is varying. Normally this would be true, but it is important to be aware of the type of processes acting on your analytical signal, especially if you are stretching your instrument and system to the limit of its specified performance.

4.3.6. 'Aliasing'

Some very peculiar effects can occur as a result of the discrete sampling performed in the conversion of analogue into digital signals.

∏   Look at the diagram in Fig. 4.3c. You can see that a high-frequency signal has been sampled at regular intervals.

Draw a smooth low-frequency curve through the voltages at each of the sample points. These voltages have been indicated by a dot at each sample point on the diagram.

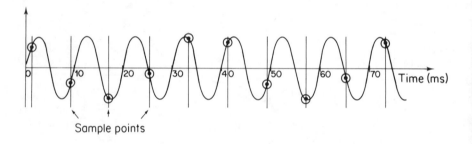

Fig. 4.3c

Measure:

(*i*)   the frequency of the original signal,

(*ii*) the sampling rate,

(*iii*) the frequency of the 'apparent' lower-frequency wave.

The simplest wave connecting each of the sample voltages has been drawn in Fig. 4.3d, together with the periods t_h, t_s and t_l.

(*i*)   Period of the high-frequency signal, t_h = 10 ms. Thus the frequency = 100 Hz.

(*ii*) The time between samples, t_s = 8.0 ms. This gives a sampling rate of about 120 Hz.

(*iii*) Period of low-frequency wave, t_l = 40 ms. Thus the frequency = 25 Hz.

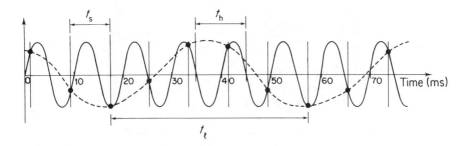

Fig. 4.3d.   *Aliasing*

If you have difficulty with (*i*) or (*ii*) then you should refer back to Section 2.1.9.

We can now see that if you take a series of 'snapshots' of a wave then it is possible to get a series of voltages that will actually correspond to a wave of a completely different frequency. This is called '*aliasing*'. The importance of introducing this here is to illustrate that, under certain circumstances, your computing system may well be 'seeing' something very different from what you imagine.

SAQ 4.3a   We wish to convert the chromatogram signal, sketched in Fig. 4.3e, into an 8-bit digital signal, the only ATD converter available having an input range of 1.0 V.

(*i*)   Calculate the possible percentage error due to conversion in recording each peak.

$\longrightarrow$

SAQ 4.3a
(cont.)

(*ii*) What would be required of an amplifier if
we were to use it to reduce the errors arising
because of quantisation?

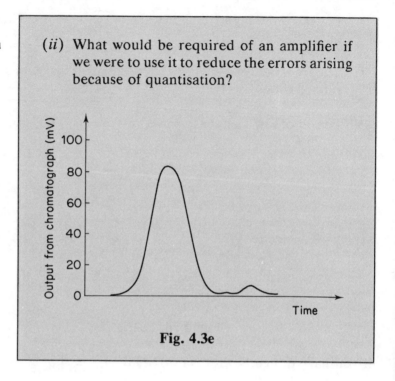

Fig. 4.3e

SAQ 4.3b | Explain what would be the effect on the electrocardiogram (heart-beat signal) given in Fig. 3.1n if it were fed (with correct amplitude) into a DTA converter with a sampling range of 100 Hz.

4.4. PRESENTATION OF DATA

4.4.1. Introduction

There are very many ways in which data can be transferred from the instrument to the outside world. Of course, with the development of computers, that 'outside world' often consists of some form of intelligent electronic system. However, much raw data is still presented in traditional forms, eg on meter display or on chart recorders.

In this section we shall consider mainly the traditional systems and refer the reader to the specialist unit on microprocessors for more advanced data-handling. We shall use the chart recorder and digital display as specific examples to illustrate some general principles.

4.4.2. Output from Instruments

We first introduced the idea of the Read-Out System of an instrument in Section 1.2.8. The output information may be displayed directly by the instrument using a built-in display system. For example, the instrument may have its own output meter, digital display, or even its own chart recorder. Alternatively the information may be available in the form of an electrical signal from output terminals. In this section we shall concentrate on the treatment of the situation when the output is in the form of an analogue voltage signal.

It is also common now for the electrical signal to be in a digital form, either in serial (eg RS 232) or parallel (eg IEE 488) format. For this type of interfacing the student must refer to the unit on microprocessor applications.

There are two important factors that must be known about the analogue output of an instrument:

(*a*) Maximum output-voltage (or voltage range), V_{om},
(*b*) Output impedance, R_o.

The voltage range differs considerably from instrument to instrument. For example, one instrument may give a maximum output of 10 mV, whereas another may give up to 10V. Moreover, two output ranges may be available from the same instrument. We have already met the concept of output impedance in Section 2.3.4, in which it was shown that a source with a high output impedance is capable of giving only a very small current. For an ideal measurement of voltage the output impedance of the source should be much lower than the input impedance of the recording instrument.

∏ A spectrophotometer has two sets of output terminals for connections to a printer and a chart recorder respectively as in Fig. 4.4a. If the voltage range, V_{op}, for the printer output is 1.00 V, calculate the voltage range, V_{or}, available at the output terminals for the recorder.

Hint: Use the concept of a potential divider.

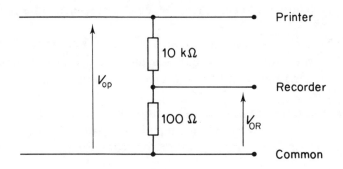

Fig. 4.4a. *Potential-Divider output circuit*

The circuit given in the diagram is similar to the potential divider in Fig. 2.1f with,

$$R_2 = 10 \text{ kohm} \quad \text{and} \quad R_1 = 100 \text{ ohm.}$$

By comparison with the equations in Section 2.1.6 we obtain,

$$V_{or} = V_{op} \times 100/(100 + 10000)$$

Thus

$$V_{or} = 1.00 \times 100/10100 = 9.90 \times 10^{-3}$$

$$= 9.90 \text{ mV}$$

This is actually referred to as a '10 mV' output. Given the limited accuracy of the resistors used in this particular equipment, this is a reasonable approximation.

If you find it difficult to remember the workings of the potential divider then you should refresh your memory by referring back to Section 2.1.6.

In general we find that the analogue output from an instrument can represent a signal in several different ways:

(*a*) a constant (DC) reading, eg the output from a pH-meter,

(*b*) a signal that varies with time, and for which 'time' is the important system variable, ie the output from a gas chromatograph, or an electrocardiogram,

(*c*) a signal that varies with time but which is simultaneously correlated with another system variable, eg the output from a recording spectrophotometer.

4.4.3. General Specifications of a Display Instrument

We have mentioned that the display system may be incorporated into the analytical instrument itself, or it may be connected to the instrument *via* output terminals. In either case we are normally considering a system such as a meter,which accepts an input voltage, V, and produces some form of visible response, R.

System	Response
Moving-Coil Meter	Deflection of pointer against a scale
Digital Display	Change in display digits
Chart Recorder	Deflection of writing pen.

There are three important criteria to describe the performance of a recording system; sensitivity, response-time, and input-impedance. These are introduced below.

The *Sensitivity*, S, of the instrument may be defined as follows:

$$S = \frac{\text{Output Response}}{\text{Input Signal}}$$

This will obviously have various units depending on how the response is defined.

It is also quite common for the sensitivity of a display system to be expressed in terms of the Input-Voltage required to give a Full Scale Deflection (FSD). For example, the sensitivity of a particular range of a chart recorder may be given by saying that the input voltage range is 10 mV. This means that the pen will move from one side of the paper to the other for a change of input voltage from 0 to 10 mV.

It is also common for instruments such as a chart recorder to have the facility of changing the sensitivity with a Range Switch giving various input ranges: eg 10, 20, 50, 100, 200, 500 mV and 1, 2, 5, 10 V.

∏   A chart recorder has a chart width of 20.0 cm. If a given input voltage produces a deflection of 1.7 cm when the input range setting is 100 mV, calculate what deflection would be produced by the same voltage if the input range is changed to 1.0 V.

On changing from an input range of 100 mV to one of 1.0 V the sensitivity of the recorder is being *reduced* by a factor of ten. It now requires 1000 mV to give the same deflection (FSD) that was previously given by 100 mV. If the sensitivity is reduced by a factor of ten, then the output goes down also by a factor of ten, ie from 1.7 cm to 0.17 cm. The answer is 0.17 cm.

The *Response-Time* of a display system gives an indication of the time it takes to respond to a given input-signal. We saw in Section 3.3.4 and 3.5.4 that this is an important consideration if we are to ensure that the system does not attenuate any of the Fourier components of the signal, and if we wish to reduce high-frequency Fourier components of the noise spectrum.

The importance of *Input Impedance* in any measuring system was discussed in Section 2.3.6 and we do not propose to spend much time on this topic in this section. The input impedance of chart recorders may vary considerably, sometimes depending on the range being used. For example, one particular recorder has an input impedance of 100 Mohm on its millivolt ranges and 2 Mohm on its volt ranges. The Input Impedance of most digital voltmeters is usually at least 10 Mohm. With this very high value there is rarely any significant error in reading voltage when the voltmeter is connected to most instrument output terminals.

4.4.4. Chart Recorder

The flat-bed chart recorder is a common instrument in most laboratories. The term 'flat-bed' refers to the fact that the chart paper is drawn out over a flat surface (or bed) as opposed to types which use a cylindrical drum as the supporting surface.

There are two main modes of use. A *Y-t* recorder has a single input which deflects the pen on the *Y*-axis while the paper moves at a constant speed perpendicular to this *Y*-deflection. This allows the signal variable to be plotted as a function of time. The *X–Y* recorder has two signal inputs. One deflects the pen along the *Y*-axis (as in the *Y-t* mode), but the other voltage input deflects the pen on a perpendicular *X*-axis. The paper does not move. This allows the variation of one variable (*X*) to be plotted as a function of another variable (*Y*). It is possible for certain recorders to be used in either of the two modes.

The sensitivity of the recorder can usually be changed by using a built-in amplifier or attenuator so that it is possible for different voltage ranges to give the same deflection (FSD) on the output.

∏   A particular chart recorder has two input sensitivity controls. One switches between ranges 1, 2, 5, 10, 20, and 50 mV. The other is a two-way switch labelled × 1 and × 100. The FSD is 20.0 cm.

Calculate the input voltage required to give a deflection of 1.50 cm with a setting of 20 mV and × 100.

You may well have been confused about the exact meaning of the '× 100' factor. Does this represent an increase in sensitivity by a factor of 100 or an increase in the output range? This is actually an example of poor equipment labelling which can lead to errors in measurement. Hopefully you will have reasoned that since the input controls are expressed in terms of 'range' then the 'x100' gives an increase in *range*.

Thus the actual input range becomes,

$20.0 \text{ mV} \times 100 = 2.00 \times 10^3 \text{ mV}$ for a full scale deflection.

Hence,

20.0 cm (FSD) is equivalent to 2.00×10^3 mV

1.50 cm deflection is equivalent to $(1.50/20.0) \times 2000 = 150$ mV

The correct answer is 150 mV.

If you incorrectly assumed that ' $\times 100$' represented an increase in sensitivity then this will *reduce* the input range, ie 20.0 mV/100. Thus 0.200 mV would then be equivalent to 20.0 cm (FSD), giving 1.50 cm deflection as equivalent to 0.015 mV (wrong answer).

The time taken for a chart recorder to respond to a given step in input can be defined in a number of ways depending on the characteristics of the system. The *Response-Time*, can be defined as the time it takes for the pen to move from 0% to 90% of the total deflection in response to a sudden change in the voltage input. Compare this with the *Time-Constant* for an amplifier given in Section 3.2.5.

For a mechanical system such as a chart recorder or a moving coil meter, the response-time cannot usually be much less than about 0.5 s. Some specific mechanical systems are made with shorter response times particularly for use in biological or physiological applications in which periods of less than 0.1 s are important, eg an electrocardiograph machine.

It is also common practice to give the *Maximum Writing-Speed* of the pen. This gives the maximum speed at which the pen can move across the paper in response to a step-voltage input. Note that this refers to the movement of the pen along a voltage axis (X or Y) and not along a time (t) axis.

∏   Refer to Fig. 4.4b, which shows the response time of a Y-t chart recorder to a step input voltage. The speed of travel of the paper along the t axis is 6.0 cm s^{-1}.

Calculate:

(*i*) the Maximum Writing-Speed of the pen across the paper,

(*ii*) the Response-Time of the system.

Estimate the approximate high-frequency cut-off, f_h, for this system.

Hint: refer back to Section 3.2.5.

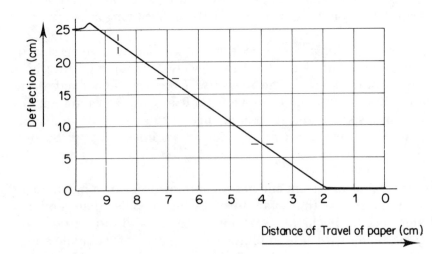

Fig. 4.4b

(*i*) The Maximum Writing-Speed is applied only to the speed of pen travel along the Y-axis and can be calculated by measuring how long it takes the pen to move between two positions in the centre of its travel.

We choose to measure the time it takes to go from the 7.0 cm(Y) mark to the 17.0 cm(Y) mark. This is equivalent to 3.0 cm (t) on the time axis, which is then equivalent to $3.0/6.0 = 0.50$ s. Thus the maximum writing speed is $(17.0 - 7.0)/0.50 = 20$ cm s^{-1}.

If you have 20.88 cm s^{-1} for an answer then you may have calculated the speed of the pen *over the paper*, which also takes into account the speed at which the paper is travelling. In this problem we are only concerned with the physical speed (in the Y-direction) of the pen.

(*ii*) If we define the Response-Time as being the time taken for the pen to travel from 0 to 90% of its final value, then we must measure for the pen moving from 0.0 cm(Y) to 22.5 cm(Y) (90% of 25.0 cm). On the time axis this is 6.7 cm, equivalent to a time of 1.1 s.

If we wish to estimate the high frequency cut-off for this recorder then we can use the relationship given in Section 3.2.5, ie

$$f_h \simeq 1/2.7\,t$$

where t is the response time.

This gives us a value

$$f_h \simeq 1/3.0 \simeq 0.33 \text{ Hz.}$$

This is only a very approximate value, but it does help the operator if he has an approximate idea of the order of magnitude of the frequency response of his equipment.

Most chart recorders also have a Zero-Offset Control. This has the same effect as the zero-offset control in other DC amplifiers (see Section 3.2.3) and usually enables the operator to set the zero anywhere over a large range. It is even possible to set the zero outside the range of the pen movement. This allows the operator to compensate for the presence of a large and constant DC voltage which may be mixed with the smaller varying analytical signal voltage.

A chart recorder is essentially a DC system, and, as such, will be susceptible to drift. However, the Zero-Offset Control can be used to compensate for this and it is important, when using a chart recorder, always to check the setting of the zero. The zero position can be set by *short-circuiting* the input terminals (either by an external connection or by a 'zero' switch on the instrument) and then adjusting the Zero-Offset Control to get the pen to the correct position.

∏   A chart recorder has an input-range setting of 100 mV and a pen-range of 20.0 cm.

         With a voltage, V, connected to the input, the position of the pen is 17.0 cm, but when the input terminals are shorted together the pen moves to the 5.0 cm position. Calculate the voltage, V.

The input range is 100 mV and the pen range 20.0 cm. Thus the sensitivity can be expressed as $100/20 = 5.0$ mV cm^{-1}.

The important part of this problem is to realise that a zero voltage input does not correspond to the 0 cm mark on the paper.

We are told that when the input terminals are shorted together the pen is at the 5.0 cm position. In this state the voltage input to the recorder is 0 V due to the short circuit. Thus 0 V is equivalent to the 5.0 cm mark. We are also told that the unknown voltage, V, produces a pen position of 17.0 cm. The magnitude of this voltage $= (V - 0)$ V. This is therefore equivalent to $17.0 - 5.0$ cm on the paper.

The voltage, V, is equivalent to a pen movement of 12.0 cm. This is then equivalent to a voltage of

$$12.0 \times 5.0 = 60 \text{ mV}$$

Some recorders also have the facility to switch between 'variable' (var) and 'calibrated' (cal) modes.

In the calibrated mode the response of the recorder is a calibrated function of the input, eg for an input voltage range of V volts (eg 10 mV) the pen will move a full scale deflection for an input voltage of V volts. In the variable mode, pen response can be adjusted for any given voltage input. This option can be used to alter the sensitivity of the recorder to suit some particular purpose. If you wish to use the calibrated scale of the instrument then you must remember to check that it is indeed switched to the 'cal' position.

4.4.5. Digital Display

Most digital display systems that you will meet are built into the instrument concerned. However, you may decide to connect a digital voltmeter (DVM) to the analogue output socket. The sensitivity of a digital display system is defined by the input voltage required to give the maximum display value. The size of the actual display is usually given in terms of the number of digits in the display.

It is very common for a display description to include a 'half-digit', eg $3\frac{1}{2}$ or $5\frac{1}{2}$ digits. The 'half-digit' refers to the first digit in the number which is either '1' or left blank. The reason for this is that provided that this digit has a possible value only of '1', a negative sign can be incorporated along with the first digit. This means that it is not necessary to include an extra display sector just to display the sign. Thus for a display with 4 sectors which also incorporates a negative sign (ie a $3\frac{1}{2}$ digit display) the range is from -1999 through 000 to 1999.

The actual display is driven by using electronic silicon chips operating on digital electronics. Thus it is necesssary to convert the analogue voltage input into a digital signal by using an ATD converter, see Fig. 4.4c.

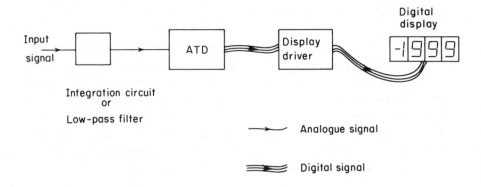

Fig. 4.4c.   *Conversion of a signal for digital display*

We saw in Section 4.3.4 that an ATD conversion takes a finite time to occur, the *Conversion Time*. However, although this time may be quite long in electronics terms (up to several milliseconds) it is still very short compared to the ability of the eye to perceive changes in the display. The digits in the display can, for a rapidly changing signal, change more quickly than the eye can follow. It is possible for a signal which appears stable when displayed on a chart recorder to give quite erratic readings when fed into a digital display.

∏   A DC signal of magnitude 10 mV suffers from noise at 100 Hz (due to poor rectification of a mains signal) with a peak-to-peak amplitude of 2 mV. Describe how the signal will appear if it is presented to the operator through:

   (*i*)   a chart recorder with a response time of 0.5 s.

   (*ii*) a digital display which samples the signal with a 'sample-and-hold' circuit approximately every 8 ms. (Hint; refer back to Section 4.3.6.)

(i) If the response time of the recorder is 0.5 s, then by using the same method as prviously in this section, we find that the high frequency cut-off, f_h, for this recorder is about 0.7 Hz. This means that a signal with a frequency much greater than this value will be severely attenuated.

The DC signal has frequency components near 0 Hz and as such will not be attenuated by the recorder. However, the 100 Hz noise is beyond the cut-off point of the recorder and will be almost totally eliminated. The mechanical system of the recorder will not be able to respond fast enough to draw out the 100 Hz noise. Thus the trace drawn on the paper will correspond correctly to the 10 mV DC signal and will have very little noise.

(ii) This situation closely resembles the one given in Section 4.3.6. In that example we found that the output from the sample-and-hold circuit still gave a variation in the value being measured, and that this variation still had the same peak-to-peak value (2 mV in this question) as the original 100 Hz noise. If this is put onto the digital display then we shall see the digital value following the low-frequency variation drawn in the answer. It may not be possible for the eye to follow exactly what is happening as the digits fluctuate, but the observer may see values between 9 mV and 11 mV fluctuating on the display, equivalent to 2 mV noise superimposed on the 10 mV DC signal.

It can be seen from the exercise that, because the sample-and-hold circuit records the value of the signal over a very short time, the digital display system can be affected by high-frequency noise due to the spurious values for the signal that may occur just at the instant of sampling.

∏ Refer back to the previous problem. What type of circuit do we need to introduce *before* the ATD converter if we wish to obtain a digital display with the same stability as the record on the chart recorder?

You may have chosen one of two possible answers: either a *low-pass filter* or an *integrating circuit* The purpose of the circuit used must be to eliminate high-frequency noise. Both of these circuits will do this.

If an integrating circuit is used, an integration time of 1 s (say) may be chosen, and then the system will integrate the signal for 1 s, then display the result as a constant digital value. The system will then repeat the process, updating the value of the digital display every second. If a low-pass filter circuit is used the digital display immediately records the output value of the filter and will change accordingly. However, because of the smoothing effect of the filter the display would not now be so erratic.

We see from this text question that we can use either a low-pass filter (damping) or an integrating circuit to reduce the effect of high-frequency noise on the digital display. The concept of the response-time of the digital-display system will then depend on the type of smoothing circuit used. It will be given by a Time-Constant if a low-pass filter is being used, or by the Integration-Time for an integrating system. Each system has its own advantages and disadvantages. The integrating system may be preferred for displaying a non-varying signal because its inherent stability during the integration period makes it easy to read.

However, when setting-up an instrument it is often necessary to adjust some variable parameter (eg wavelength) and observe the changing output reading until a maximum value is obtained. With an integrating circuit the output reading will not change at all until the end of the integrating period making it difficult to see immediately whether the adjustment is being made in the correct direction. With the filter circuit, although the output may be damped, it would still show in which direction the reading is changing.

SAQ 4.4a | A chart recorder has possible voltage input ranges of 10.0 mV, 100 mV and 1.00 V, and a chart width of 20.0 cm.

Express the maximum sensitivity of this recorder in units of cm mV^{-1}.

SAQ 4.4b | A $5\frac{1}{2}$ digit display system has an input voltage range of -2.0 V to $+2.0$ V.

What change of input voltage will cause the least significant digit to change by one unit?

SAQ 4.4b

Summary

This part deals with the performance of some of the functional elements in instruments which affect the interpretation, presentation, or passage of data.

The first section shows the way in which the mathematical concepts of integration and differentation are used in instrumentation. By using the Fourier Transform concept developed in Part 3, the benefits and dangers of these techniques are highlighted.

The second section deals with instruments which use information in the form of pulses, concentrating on the possible limitations and errors which may occur.

Pulses of a different form are introduced in the third section which deals with digital signals. Emphasis is again placed on the possible, and sometimes unexpected, errors that can occur when analogue and digital signals interconvert.

The last section discusses the problems that arise in presenting data in different forms, either in an analogue format on a chart recorder or in a digital display.

Objectives

It is expected that, on completion of this Part, the student will be able:

- to describe the uses of 'integration' in an instrumental system,

- to explain the errors that can occur when we use integration and differentiating systems in instruments,

- to explain how a multichannel analyser can provide a 'spectrum' of high energy radiation,

- to explain the statistical error in counting random pulses,

- to calculate the quantisation error of conversion in expressing an analogue signal in a digital form,

- to describe the effect of sampling-rate and conversion-time in the conversion of an analogue signal into a digital signal,

- predict the performance of a chart recorder from its specification,

- predict the performance of a digital display from its specification.

5. Complete Instrument Systems

Overview

The final part brings together the topics introduced earlier and relates these to the final design and specification of instruments. Some spectrophotometer systems are used to illustrate the applications of fundamental concepts of instrumentation. However, readers who wish to study spectrophotometers in depth must refer to the specialist Units on that topic.

5.1. SPECTROPHOTOMETER SYSTEMS

5.1.1. Introduction

In this section we choose to study some spectrophotometer systems that have been developed from the humble colorimeter which was introduced earlier in this unit. Spectrophotometers as complete instruments will also be discussed in more detail in the special units on spectroscopy. However, it is useful to use these systems as very good examples of certain important aspects of instrument design.

The first and most important difference in sophistication between a colorimeter and a spectrophotometer (single or double-beam) is in the method of isolating particular wavelengths. The colorimeter uses coloured filters (hence the name *color*imeter) to select different bands of wavelengths. The choice of bands or wavelengths depends on the available filters. With a spectrophotometer, however, all the wavelengths within the range of the instrument, can be scanned thereby allowing the operator to select any wavelength over the relevant spectral range (hence the name *spectro*photometer). There is thus no need for the operator to introduce filters.

Dispersive spectrophotometers use an optical device known as a monochromator to select the wavelengths. The term dispersion describes how the radiation from the source is *separated* out into different wavelengths. Only a selected narrow band of wavelengths is used at any given instant. The instrument can be made to scan the different wavelengths, either by manual control or by an automatic sweep-mechanism.

5.1.2. Double-beam Systems

A *double-beam* spectrophotometer is similar to a colorimeter in that it can be used measure the absorption, by the sample, of a selected wavelength of the em spectrum. However, in addition to having more sophisticated component parts, the double-beam spectrophotometer also has a *fundamentally* more accurate method of making the measurement.

Single-beam spectrophotometers use a measurement method similar to that of the colorimeter, and they both suffer from the same inherent disadvantage. All single-beam instruments have to be calibrated by using a 'blank' solution and the sensitivity must be adjusted to give 100% T (or 0A) before the absorption of the sample is measured. (T = transmittance, A = optical absorbance). The output from the lamp and the sensitivity of the detector vary with wavelength, hence the 'blank' calibration for 100% T must be done whenever the wavelength is changed. This is time-consuming and, as we shall see below, may allow serious errors to creep into the measurement.

∏   A single-beam spectrophotometer has a lamp the output from which varies in intensity. If the intensity of the em radiation from the lamp of such a spectrophotometer decreases after the 'blank' calibration has been performed, and before the sample measurement is made, will the *apparent* absorbance, A, of the em radiation be:

(*i*)   greater than the *true* value,

(*ii*)   less than the *true* value,

(*iii*) the same as the *true* value?

NB Remember that the absorbance, A, of a sample increases if its transmittance T, decreases, and *vice versa*.

If the intensity of the light source decreases after calibration of the instrument, the *effect at the detector* is the same as if the sample had absorbed more light. Thus if we read the absorbance of the sample on the meter of the instrument, not knowing of the decrease of light intensity, we shall think that the absorbance, A, of the sample is greater than its true value. Drift in the source-intensity of a single-beam spectrophotometer creates a direct error in the reading of the instrument. The answer is (*i*).

The type of error mentioned above may arise also as the result of DC drift in the optics, the electronics, or the detector of the instrument. In fact, we have already seen in Section 3.2.3 that drift can be a problem with any DC system, but how can it be avoided here?

We now describe the basic optical double-beam system as shown in Fig. 5.1a. The beams may be ultra-violet, visible, or infa-red radiation, depending on the type of spectroscopy. The output from the source is split into two beams S and R which travel through the *sample* and the *blank* solution respectively. The blank solution is now referred to as the *reference solution*.

As shown in the diagram each of these beams is directed to the detector. The mirror, C, is a rotating disc with a profile as shown in the diagram. It consists of six segments, two of which reflect the

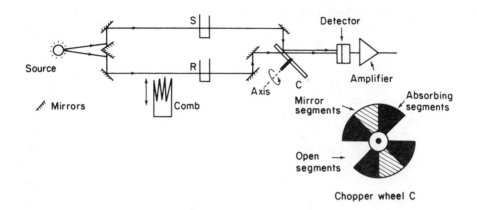

Fig. 5.1a. *Double-beam optical system*

R beam into the detector and two of which are cut away to allow the S beam to reach the detector. Between each open and reflecting segment there is an absorbing segment which prevents any radiation from reaching the detector. The disc is positioned with its axis just away from the beams. As it rotates the various segments allow the S and the R beam to reach the detector alternately with a brief period of total absorption (darkness) between each beam. The rotating mirror, C, is called a 'Chopper' mirror for obvious reasons.

∏   Refer to Fig. 5.1a and assume that either beam (S or R) would cause a voltage of 1.0 V at the output of the amplifier, provided that neither beam was attenuated in any way.

Decide which of the voltage waveforms, (i), (ii), (iii), or (iv), shown in Fig. 5.1b would be generated at the output of the amplifier if,

the sample has a transmittance of 40%,

the reference beam is not attenuated in any way, and

the chopper mirror rotates at 12.5 revolutions s^{-1}.

Indicate your choice below

(*i*)   right / wrong

(*ii*)   right / wrong

(*iii*)   right / wrong

(*iv*)   right / wrong

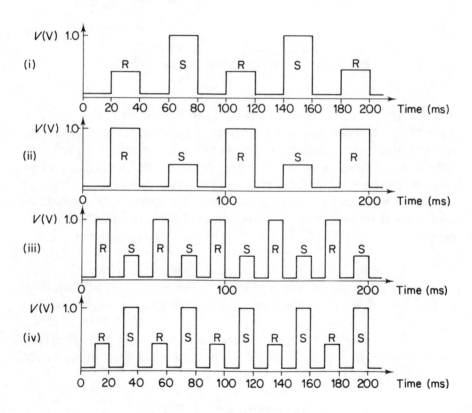

Fig. 5.1b

The answer is:

(*i*)   wrong,

(*ii*)   wrong,

(*iii*)   right,

(*iv*)   wrong.

The reference beam, R, is not attenuated, and will therefore give the maximum value, 1.0 V, at the output. The sample beam, S, is attenuated and thus will give a *lower* output voltage from the detector. Obviously the waveforms (*i*) and (*iv*) are incorrect because the voltages for S are greater than for R.

The waveform (*ii*) and waveform (*i*) again, are wrong because of the period of the waveform. The chopper is rotating at 12.5 revolutions s^{-1}, and this is equivalent to a period for each cycle of 1/12.5 s = 80 ms. However, you can see from the diagram, Fig. 5.1a, that in each revolution there will be two occasions when each of the R and S beams are able to reach the detector. Thus within a period of 80 ms there must be two pulses from each of the beam. The waveforms (*i*) and (*ii*) contain only one pulse from each beam in this period.

The waveform (*iii*) is the only one which satisfies all the requirements. You might also notice that the voltage level for S is 0.4 V. The voltages for S and R are in the same ratio as their light intensities, and thus the response of the detector must be *linear*.

We see that absorbance in the sample causes a difference between the height of S and R in the signal from the detector.

If we can now process this information properly we should be able to get a signal in the instrument which is directly related to the transmittance, T, or absorbance, A, of the sample. Before we see how this is done work through the next exercise.

∏   What is the main advantage of a double-beam system with
       a 'chopper' mirror over the single-beam system used in the
       colorimeter?

       Hint: refer to the previous exercise in this section.

Refer back to the earlier exercise, where we found that, in a single-
beam spectrophotometer, there is a danger that DC drift between
calibration with the 'blank' and taking the sample measurement may
cause an error in the result.

A double-beam system, however, is continually switching between
the sample and the reference (or blank) solution. In effect, it is re-
calibrating itself every time the chopper switches between the two
beams. It reduces the time between calibration and measurement
from minutes to fractions of a second. Slow drifts due to $1/f$ noise
(see Section 3.5.3) with periods greater than the period of rotation of
the 'chopper' cannot now affect the result. The process of 'chopping'
the optical beam is a form of *modulation* as discussed in Section 3.4.

You should have been able to anticipate that the system with the
'chopper' mirror would have had the effect of reducing errors due
to drift. We saw in Section 3.4.2 how *modulating* the DC signal allowed us to reduce the drift occurring in a DC amplifying
system.

Having decided that a double-beam system has inherent advantages
over a single-beam system, we now look at how it can be used in
an instrument. How can we convert the signal in Fig. 5.1b(*iii*) into
a value for the %*T* or *A* of the sample? There are two principal
methods: *optical-null*, and *ratio-recording*. Because of advances in
electronics most new instruments are of the ratio-recording type but
we first describe the optical-null type because there are still many
instruments with that system being used.

5.1.3. Optical-null Balance

Optical-null is, at its name suggests, a null method (see Section

2.4.3). The absorbance in the sample beam, S, is balanced by an equal, but *artificial* 'absorbance' in the reference beam. This artificial absorbance in the reference beam is achieved by sliding a carefully manufactured 'comb' (Fig. 5.1a) into the R beam. The comb blocks out a varying proportion of the beam until it exactly balances the absorbance due to the sample in the S beam. This null-point is detected when S and R are the same height in the waveform in the signal coming from the detector system (cf Fig. 5.1b(*iii*)). The position of the comb across the R beam will depend on how much of the beam must be blocked to balance the absorption in the sample beam. If the absorbance of the sample is large, then to obtain a balance, the comb must move to block the majority of the R beam. The *position* of the comb is therefore directly related to the *absorbance* of the sample. The attenuating comb is usually connected directly to the pen of a chart recorder and the movement of the comb gives a direct reading on the chart paper of the change in absorbance (or transmittance) of the sample.

∏   Refer to the waveform (*iii*) in Fig. 5.1b, which is produced at the output of the detecting system when the sample has a transmission of 40%.

   (*i*)   By using the concept of a Fourier Transform, calculate the *two* lowest AC frequencies in the waveform.

   (*ii*) Which of the two frequency components exists *only* if there is a difference in intensity between S and R?

   (*iii*) What information is carried by the other of the two frequency components?

   Hint: refer to Section 3.1.4.

The two lowest AC frequencies (see Fig. 3.1g) are:

$$f_1 = 25 \text{ Hz},$$

$$f_2 = 50 \text{ Hz}.$$

The higher frequency (shorter period) component arises because of the alternation between the absorbing and non-absorbing segments of the chopper. If the average magnitude of the two pulses were to decrease then the amplitude of this component would also decrease. The lower-frequency component arises from any difference between S and R. Hence if S and R are of the same magnitude then the lower-frequency component would have a zero amplitude, ie it would cease to exist.

Hence the correct answer to the question is:

(i) $f_1 = 25$ Hz, and $f_2 = 50$ Hz.

(ii) f_1 (25 Hz).

(iii) The higher-frequency component, f_2, has an amplitude proportional to the average amplitudes of the two pulses.

If you had difficulties in deriving these answers then you should refer back to the answer to Section 3.1.4 in which the same question is expressed differently.

We can see that the signal from the detector carries two separate pieces of information:

(a)   the difference in intensity between S and R (at a frequency, f_1),

(b)   the average intensity arriving at the detector (at a frequency, f_2).

These two pieces of information are carried in Fourier components at different frequencies, f_1 and f_2, respectively.

The frequency component, f_1, can be isolated with a tuned filter (Section 3.3.3) and used, *via* a phase-sensitive detector (Section 3.4.4), and an amplifier, to drive the motor which moves the attenuating comb into the reference beam, see Fig. 5.1c. The comb will move only if there is a difference between S and R, and it will move in such a direction that it alters the amount of the R beam blocked until it balances the absorbance in the S beam.

Note that the phase sensitive detector takes its reference signal (see Section 3.4.4) from the chopper wheel so that it 'knows' which beam, S or R, is being used at any given instant.

In some spectrophotometers, the frequency component, f_2, can be isolated and used to operate a system called an Automatic Gain-Control (AGC) – see Fig. 5.1c. If the average light-intensity is low the amplitude of f_2 will be small and, sensing this, the AGC will increase the *gain* of the detector to restore the voltage output to an acceptable level.

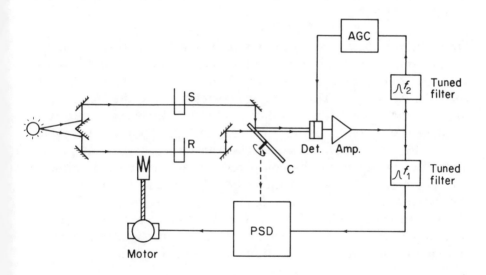

Figs. 5.1c. *Double-beam optical null system*

One severe disadvantage of the optical null system is that if the sample has a very high absorbance (S is very small) then, when the null-balance point is reached, the reference beam is also severely attenuated (R is very small). Although an AGC circuit can increase the gain of the detector, this also increases the *noise* at the detector, which ultimately limits the maximum values of absorbance which can be measured with the instrument. These spectrophotometers often have an indication of the light energy reaching the detector, with a warning when it is becoming too low.

5.1.4.   Ratio-recording

The ratio-recording instrument does not use an optical null-balance method, and there is no attenuation of the reference beam. Hence there is always a suitable amount of light reaching the detector. The signal from the detector, in the form given in Fig. 5.1b(*iii*), is taken directly and processed electronically to give the absorbance of the sample.

One method of processing the signal would be for three separate sample-and-hold circuits repeatedly to sample the R, S, and D pulse voltages from the detector (as in Fig. 5.1d) to produce three DC voltages V_R, V_S and V_D. The D voltage corresponds to the signal from the detector when all light is blocked by the 'dark' segment of the chopper. $(V_R - V_D)$ corresponds to the light passing through R. $(V_S - V_D)$ corresponds to the light transmitted by the sample, S. These two voltages may be compared on an accurate potentiometer system (cf the question in Section 2.4.5), and the position of the moving contact could be connected directly to the pen of the chart recorder, see Fig. 5.1.d. The potentiometer would be balanced automatically and continuously, by an electronic feedback system similar to that used in a potentiometric chart recorder, see Section 2.4.5.

∏   A detector in a *ratio-recording system* develops a fault which means that its response becomes non-linear, ie the voltage output from the detector is no longer proportional to the intensity of light falling on the detector. The form of this response is shown in Fig. 5.1e.

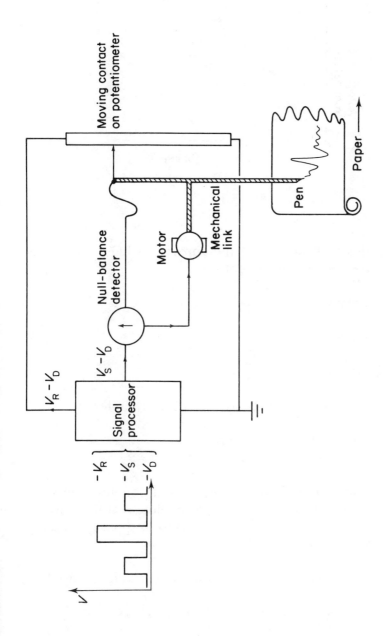

Fig. 5.1d. *Ratio-recording system*

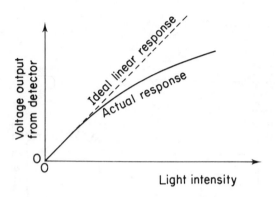

Fig. 5.1e

Will this fault mean that the apparent value of absorbance recorded by the instrument is:

(*i*)   greater than,

(*ii*)   equal to,

(*iii*)   less than,

the true value of the absorbance of the sample?

Indicate which option is correct.

The form of the output from a detector in a double-beam system is given by the waveform (*iii*) in Fig. 5.1b. Note that the waveform in that example results from a *linear* detector response to a sample transmittance of 40%.

For a percentage transmittance, T, and a *linear* detector, we have:

$$(V_S - V_D)/(V_R - V_D) = T/100$$

In this question the response of the detector is proportionally less at higher light intensities.

Hence,

$$(V_S - V_D)/(V_R - V_D) > T/100$$

The *apparent* percentage transmittance given by the instrument is now given by

$$T\ app\ =\ (V_S - V_D)/(V_R - V_D) \times 100$$

whence

$$T\ app\ >\ T/100 \times 100\ >\ T$$

where T is the *true* transmittance.

Thus if the apparent transmittance is greater than the true value, then the apparent absorbance is *less* than the true value.

The correct answer is (*iii*).

It is clear from the question that if the detector system of a ratio-recording instrument is non-linear there will be an error in the absorbance reading. It is for this reason that early instruments were mainly of the optical-null type, in which the linearity of the detector is not critical. However, improvements in technology have made the linearity of detection systems far more reliable, thus allowing the ratio-recording systems to become dominant over optical-null systems.

5.1.5. Infa-red Fourier Transform Spectroscopy

Consider the spectrum of ir radiation shown in Fig. 5.1f. The scale is calibrated both in frequency $f(ir)$ (Hz), and in wavenumbers, $\bar{\nu}$ (cm^{-1}). Wavenumber is a unit which is proportional to frequency, and is defined as the reciprocal of the wavelength in centimetres. The relationship between frequency and wavenumber is given below.

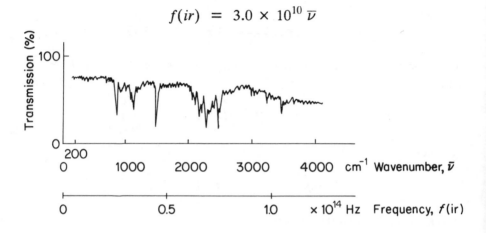

$$f(ir) = 3.0 \times 10^{10}\ \bar{\nu}$$

Fig. 5.1f.   *IR spectrum*

A 'normal' wavelength-scanning spectrophotometer will use a monochromator to isolate a very narrow band of wavelengths. It records the transmission at these wavelengths, and then derives the whole spectrum by slowly sweeping the wavelength setting throughout the range. A Fourier-Transform spectrophotometer does *not* isolate any wavelengths, and has no need of a monochromator.

The 'heart' of many Fourier-Transform spectrophotometers is an optical system called a *Michelson Interferometer* (MI) see Fig. 5.1g. It includes a 'beam splitter' which splits the incoming ir radiation into two beams of approximately equal intensity, and then allows them to recombine at the detector. There is a fixed mirror to reflect one beam, and a moving mirror to reflect the other beam.

By using the phenomenon of interference, and moving the mirror at constant speed, v mm s^{-1}, the MI is capable of 'modulating' the ir signal. It produces a new signal of frequency, f_m, which carries the same information as the original ir signal, see Section 3.4.5.

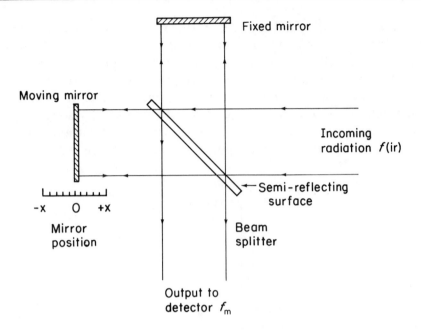

Fig. 5.1g. *Michelson interferometer*

If a monochromatic ir wave of frequency, $f(ir)$, enters the MI, then the output will be of much *lower* frequency, f_m, where,

$$f_m = \frac{v}{1.5 \times 10^{11}} \times f(ir)$$

and v is the speed of travel of the mirror in mm s^{-1}.

Note: In earlier examples of modulation in this text (Section 3.4), the information was transferred to a wave of higher frequency. However, it is possible, as in this case, that 'modulation' may transfer information to a lower frequency.

We can describe the position of the moving mirror, either in terms of its physical position (in mm), or on a real-time scale (in seconds). Obviously the time scale and position scale are related to each other by the velocity of the mirror. It is also convenient to choose both scales to be 'zero' at the centre (symmetrical) configuration of the system, see Fig. 5.1g.

A continuous spectrum, as in Fig. 5.1h(i), has components at all frequencies in the working range. Normally, when these emerge from the MI they will be in random phases with respect to each other, and the algebraic sum of all of the frequency components will then be zero. However, in the middle of the range of travel of the mirror, a perfectly symmetrical configuration of the optical system occurs, and, at that position only, all frequency components are in the correct phase, resulting in an output to the detector, see Fig. 5.1h(ii).

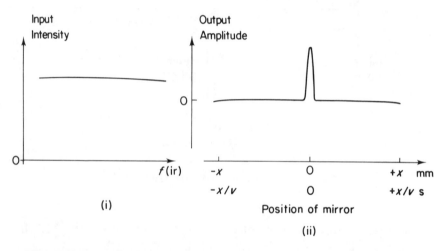

Fig. 5.1h.   *Michelson Interferometer – continuous spectrum*

The band-width of this output, as a function of the position of the mirror, both in space and time, is extremely small.

If a very narrow band of frequencies enters the MI, as in Fig. 5.1i(i), the output will be as shown in Fig. 5.1i(ii). There is an oscillating signal of frequency, f_m, which has been derived from the frequency of the incoming signal, $f(ir)$.

However, due to the line width, Δf, of the incoming signal there is more than one frequency component. Away from the centre of symmetry of the system, these components become out of phase with each other, and their combined sum averages to zero, see Fig. 5.1i(ii). The time taken for this to happen, Δt, depends on the range of frequencies, Δf.

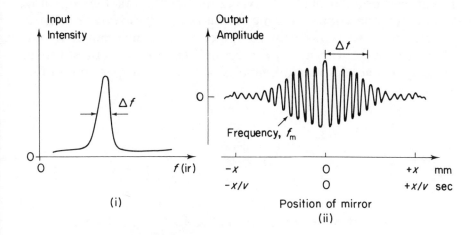

Fig. 5.1i. *Michelson Interferometer – single 'line'*

$$\Delta t \simeq 0.5/\Delta f$$

Does this equation remind you of anything? Refer to Section 3.1.6.

The oscillation frequency, f_m, carries the information giving the ir *frequency*, $f(ir)$, of the original 'line' in the ir spectrum, and the shape of the envelope of the oscillation decay gives information about the *shape* and *line width* of this 'line'.

The signal in Fig. 5.1i(ii) is a Fourier Transform of the ir spectrum in Fig. 5.1i(i). For a complete spectrum, as in Fig. 5.1j(i), the situation is naturally more complex because the Fourier Transform will contain overlapping information from all of the 'lines'. Nevertheless, the Fourier Transform (Fig. 5.1j(ii)) carries *all* the necessary information concerning the transmission characteristics of the sample.

What is the use of a Fourier Transform if we are only familiar only with the interpretation of the 'normal' spectrum?

The answer would probably be 'very little', except that there are now modern computational techniques known as *Fast Fourier Transforms*. These very powerful computer programmes make it possible to convert within a few seconds, the Fourier Transform in Fig. 5.1j(*ii*) back into a more familiar transmission spectrum in Fig. 5.1j(*i*).

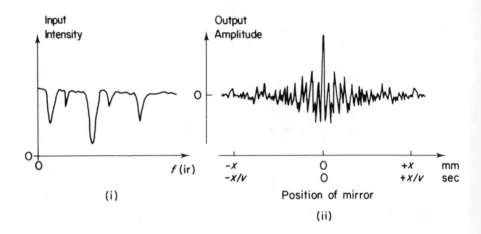

Fig. 5.1j.   *Michelson Interferometer – complete spectrum*

What are the advantages of Fourier Transform spectroscopy?

Try the following question.

∏   Complete the following sentences by using the words 'narrow' or 'wide' as appropriate.

If, in an ir transmission spectrum as in Fig. 5.1j(*i*), a particular 'line' is very narrow, then,

(*i*)   the band-width of the 'line' in the frequency domain is

        ————— ,

(*ii*) the band-width of the 'line' in the time domain is

        ————— .

(*i*)   The band-width in the frequency domain is *narrow*.

The transmission ir spectrum of a sample is already drawn in the frequency domain. The width of the 'line' in the spectrum corresponds exactly to the width in the frequency domain.

(*ii*) The band-width in the time domain is *wide*.

This follows directly by using the reciprocal relationship between signal widths in time and frequency.

Thus we can see that very sharp lines in a traditional spectrum become broad signals in the Fourier Transform.

The presence of sharp lines in a spectrum requires that a 'traditional' spectrophotometer with a monochromator must:

   (*a*) scan very slowly to avoid distorting the lines,

   (*b*) use a very narrow band-width of wavelengths from the monochromator.

Because a very narrow band only of wavelengths is being measured at any given instant, and because it will take a long time to cover the whole spectrum, the *efficiency* of making a measurement is low.

By using the MI to modulate the signal, and recording in the time domain, we convert the narrow lines into broad lines. We are also able to illuminate the sample with all wavelengths at the same time. Thus, for a spectrum with narrow lines, the Fourier Transform system allows a more efficient method of measurement to be used, and the whole spectrum is produced very much more quickly. Incidently, there are also other methods, in addition to the use of Michelson Interferometer, for obtaining the Fourier Transform of a spectrum. For more information, you are advised to consult the relevant units on spectroscopy.

The technique of Fourier Transform spectroscopy is also applied to nmr spectroscopy, where we find again that the lines in the spectrum are very narrow.

SAQ 5.1a   A spectrophotometer uses the double-beam *optical-null* principle described in the text. The detector developes a fault which means that its response becomes non-linear. The form of this response is shown in Fig. 5.1e.

Will this fault mean that when the apparent value of absorbance is recorded by the *optical-null* method in the instrument it is:

(*i*)   greater than,

(*ii*) equal to,

(*iii*) less than,

the true value of the absorbance of the sample?

Indicate which option is correct.

SAQ 5.1b

We wish to re-design a double-beam spectrophotometer so that we can *increase* the frequency of the chopper. As a result of this, which of the following statements will be true?

(*i*) The intensity of the source must be increased.

T / F

(*ii*) The intensity of the source must be decreased.

T / F

(*iii*) The response-time of the detection system must be increased.

T / F

(*iv*) The response-time of the detection system must be decreased.

T / F

5.2. COMPLETE INSTRUMENTS

5.2.1. Introduction

Near the beginning of this unit (Section 1.4.1) we discussed the difference between a straightforward *measuring* instrument and an *analytical* instrument. We suggested that the distinguishing feature of an analytical instrument was some form of experimental process being carried out within the instrument itself. However, the extent of this experimental process may well vary considerably from one type of instrument to another. The only 'experimental' process occurring in the pH meter, for example, is within the electrode system itself and, apart from changing the electrode, the operator has relatively little control over the 'experiment'. In other types of instruments, such as in chromatography and thermal analysis, both the experimental process and the need for skilled operations are a crucial factor in obtaining meaningful results.

Initially, when a new technique is developed, the analytical set-up will consist of an experiment surrounded by various bits and pieces of gadgetry! You might find it difficult at that stage to imagine how the whole system could be put into a single box. It is the commercial development of the new analytical technique which leads to the entire process being enclosed within a single instrumental entity. For example, the technique of gas chromatography was first demonstrated in the early 1950's with a relatively simple system. However, with very rapid development over a few years (because of the needs of the petrochemical industry), the gas chromatograph is now a very refined piece of technology, often with an on-board computer for control and data analysis.

If you are trying to understand a modern instrument, it might help you to try to imagine it in its original form – as an experiment surrounded by gadgets. You must get to know:

    (*a*) the experimental process being used,

    (*b*) the control available to the operator,

    (*c*) the accuracy and limitations of the instrument.

In the previous section, 5.1, we discussed the instrumentation associated with spectrophotometer systems. In this section we shall add the pH meter and gas chromatograph to the discussion, and we shall use these, together with the spectrophotometer, as examples of some modern systems.

5.2.2. The pH-meter

We have met references to both the pH meter and its electrode at various points in the previous units. We know that the pH-electrode produces a DC voltage related to the pH of the solution, and that the 'meter' is essentially a very-high-input-impedance DC amplifier, with a DC offset for standardisation.

The emf from the electrode (in volts) is given by the following equation from Section 1.3.2.

$$E_v = E_0 - 0.198 \times T \times (\text{pH} - 7.0) \ (E_v \text{ in mV}).$$

We can express the change in emf, ΔE_v, for a change of one pH unit as a function of temperature (Fig. 5.2a). See SAQ 1.3a for the appropriate calculation. The graph in Fig. 5.2a shows the *theoretical* behaviour based on the Nernst equation, and an electrode should have a response which is close to this theoretical curve. However, as an electrode ages and becomes contaminated, its sensitivity may change. We shall discuss this particular problem later.

A block diagram of a pH meter is given in Fig. 5.2.b. The Buffer Amplifier provides the very high input-impedance (see Section 3.2.6) for the meter. The Voltage Amplifier amplifies the millivolt signal to the voltage level required to give the correct reading on the meter. A DC Offset control on the DC amplifier allows us to calibrate the meter by ensuring that it reads the correct value when the electrode is placed in a standard buffer solution of known pH.

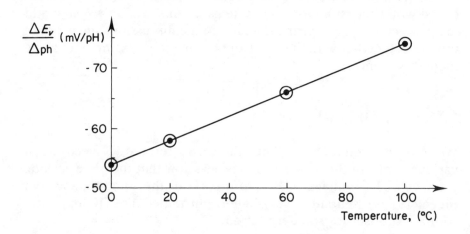

Fig. 5.2a.   *E.m.f. from a pH-electrode as a function of temperature*

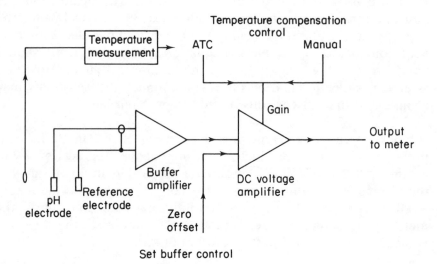

Fig. 5.2b.   *Block diagram of pH-meter*

∏ By assuming that a pH electrode at 25 °C obeys the Nernst equation, calculate the voltage-gain required in the amplifier so that the output to the meter in Fig. 5.2b changes by 0.100 V for each pH unit.

Repeat the calculation for electrode temperatures of 0 °, 25 °, 50 °, 75 °, and 100 °C, and plot your results on a graph.

At 25 °C the output of the pH electrode according to the Nernst equation should be 59 mV for each pH unit (see Fig. 5.2a and Section 1.3.2). The 59 mV signal is therefore the input to the buffer amplifier in Fig. 5.2b. The gain of the buffer amplifier is one (see Section 3.2.6), and hence the same voltage, 59 mV, is presented to the voltage amplifier. It is the function of this amplifier to increase the signal so that a signal of 0.100 V (= 100 mV) is produced at the output. In actual fact, an increase of pH by one unit gives a decrease of 59 mV from the electrode, thus our voltage amplifier must be an inverting amplifier. However, for convenience, we often omit the negative sign of G.

The gain of the amplifier, G, must be

$$100/59 \ = \ 1.69 \text{ at } 25 \text{ °C}.$$

We can repeat this calculation for different temperatures, taking the input voltage from the graph in Fig. 5.2a, for each temperature. For example at 100 °C, the input voltage is 73.9 mV. This gives a gain, G,

$$100/73.9 \ = \ 1.35 \text{ at } 100 \text{ °C}.$$

The complete graph of Gain *versus* Temperature is given in Fig. 5.2c.

We can see that some adjustment is required in the pH meter to compensate for different electrode temperatures. This adjustment must be designed to alter the voltage gain of the amplifier. This can be done manually, but it is also possible to use an ATC (Automatic Temperature Control) system which measures the temperature of the solution with a separate probe and automatically changes the gain (see Fig. 5.2b).

Most pH meters, apart from the simplest, also have a *slope* control, which can usually be set between about 85% and 110%. The 'slope' is the voltage output of the actual electrode expressed as a percentage of the theoretical output predicted by the Nernst equation, and it can be considered as a measure of the 'efficiency' of the electrode. If the pH sensitivity of the electrode drops below the theoretical value, the slope control would have to be set to a value below 100 %. For example, a slope value of 85% corresponds to a decrease of sensitivity of the electrode to 85% of the theoretical value. At a temperature of 25 °C this would mean that the change in E_v per pH unit would not be 59 mV pH^{-1} as predicted by the Nernst equation, but

$$59 \times 85/100 = 50 \text{ mV pH}^{-1}.$$

The *slope control* on the pH meter must compensate for this change in the output of the electrode so that the meter still reads the correct pH values.

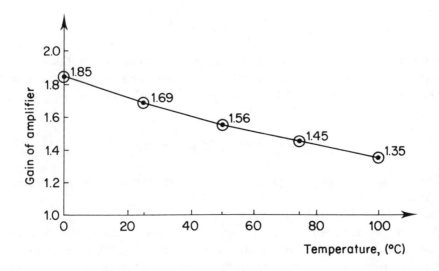

Fig. 5.2c

∏   Which of the following two statements is true?

(*i*) The *slope control* of a pH meter alters the gain of the amplifier.

T / F

(*ii*) The *slope control* of a pH meter alters the DC Offset.

T / F

From the text we can see that the slope control is required to compensate for a loss in the *sensitivity* of the electrode to pH. This is equivalent to a change in the value of A in the equation,

$$E_v = E_0 - A \times T \times (\text{pH} - 7)$$

You can see that in the equation a change of sensitivity, A, has the same sort of effect as a change in temperature, T. Hence we must compensate for changes in A in the same way as we compensate for changes in T, ie by altering the gain of the amplifier. The answer is (*i*). The slope control of a pH meter alters the gain of the amplifier.

A change in the DC offset of the pH meter has the effect of introducing a constant voltage *independent* of pH into the reading, and will *not* compensate for a change of sensitivity to pH.

The term 'slope control' is used because it compensates for a change in the gradient (or slope) of the graph of E_v as a function of pH.

We shall refer to the action of the controls of the pH-meter again in the section on calibration.

5.2.3. Gas Chromatography

It is neither possible nor appropriate to try to cover all the intricacies of gas chromatography in this section. We shall outline only the basics of the instrument in such a way as is useful to understanding of the action of the detection system.

A carrier gas is passed through a 'column', which is very often in the form of a coil with several turns, see Fig. 5.2d. The column contains packing material which supports a liquid known as the *stationary phase*. The choice of gas, liquid, and support material depends on the sample being analysed. When a sample is injected briefly into the gas, different components within the sample will take different *times* to pass through the 'column' depending on their affinity for the stationary phase. The analysis is based on the fact that a *component* can be identified by the *time* which it takes to pass through the column. There are several different detection systems available to identify different types of sample components in the gas, see, for example, Section 1.3.3. The output from the detector must be processed and fed into a chart recorder or other data-handling system.

The column coil is usually enclosed within an oven whose temperature can be made to follow a controlled variation (normally increasing) with time. The sample may also be heated as it is injected into the gas stream.

There are thus several operator-controlled functions:

(*a*) choice of gas and column materials,

(*b*) detector and injection temperature,

(*c*) oven-temperature programme,

(*d*) choice of detector,

(*e*) gas flow-rate.

Gas chromatography is still a technique in which the skill of the operator makes a very great difference to the accuracy of the result.

∏   In the diagram, Fig. 5.2d, the system is using a thermal conductivity detector (TCD) together with a separate gas by-pass, R, to the detector.

Explain the purpose of this separate by-pass path.

Hint: refer to Section 1.3.

We saw in Section 1.3.3 that a TCD works by detecting the conduction of heat away from a heated filament by the gas as it leaves the column. If the gas contains a sample then the thermal conductivity of the gas changes and we get an output from the detector. However, the transport of heat away from the filament is also sensitive to the rate of flow of the gas, and thus a variation in gas flow-rate might appear as a change in the thermal conductivity of the gas, giving a bogus analytical signal. To overcome this problem, a *differential* bridge system can be used, in which an analytical signal is produced only if the bridge becomes unbalanced (see Section 2.4.6). It is set up so that the rate of flow of gas affects both arms of the bridge in the same way, leaving the bridge balanced, and only the presence of a sample in one gas-flow causes an imbalance. The by-pass flow of gas is used to supply the second (dummy) flow of gas to the differential TCD. If the pressure of the gas supply changes, it affects both gas flows in the same way and hence does not result in an imbalance of the bridge and a bogus signal.

In the previous question, we saw that the purpose of the second gas-flow is to allow a differential measurement between the two gas-flows, only one of which contains the sample. This means that the effect of the variations in the pressure of the gas supply, and the resultant changes in gas flow-rate, will cancel out, and will not affect the output from the detector. However, in a normal GC experiment, the temperature of the oven is raised during the measurement according to some predetermined temperature programme. This change in temperature affects the properties of the gas and column, resulting in a change in the gas flow-rate through the 'column' coil. The gas flow-rate through the 'by-pass' may not change by the same amount as the flow-rate through the column, giving an imbalance, and hence an offset, in the output of the detecting system. The magnitude of this offset will change as the temperature of the oven increases.

∏ Can you suggest a way of slightly modifying the system to reduce this offset, which would be due to the inequalities between the column and the by-pass shown in Fig. 5.2d.

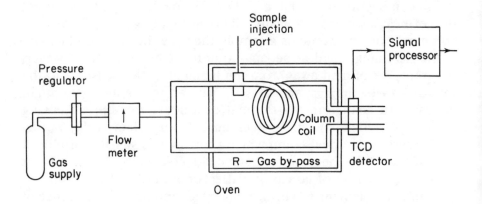

Fig. 5.2d. *Simple gas chromatograph*

The imbalance in gas-flow occurs because of the changing properties of the column and its packing materials as the temperature is raised. The only way to ensure that any changes are the same in the sample path *and* the by-pass path is to make both paths physically similar. This means that the by-pass path should, if possible, be a column of similar dimensions, and with the same packing material, as the sample column.

We have seen, in previous questions, the technique of using a double-column system to make a differential measurement. This is comparable to the use of a double-beam system in a spectrophotometer and of a differential method in thermal analysis (see SAQ 2.4a).

Π   A Flame-Ionisation Detector, FID, in a gas-chromatograph system can receive only a *single* gas-flow. Hence it cannot be used *on its own* in a differential method in the same way as a TCD. Although the FID is not as sensitive to gas flow-rate as the TCD, there is still a significant offset in the output from the detector as the temperature of the column in the oven is raised.

Can you suggest a way in which a differential FID detection system can be designed by using two columns?

A clue to the answer was given in the question. It was emphasised that the FID could not act in a differential mode *on its own*. The answer then is to use *two* FIDs, one for the sample column, and one for the dummy column. The two signal outputs from the detector will both contain a similar DC offset signal, but only the output from the sample detector will contain a signal relating to the presence of the sample. If we then process the two signals so that the dummy signal is subtracted from the sample signal we should ideally be left with only the net sample signal.

This method of using two detectors is not as satisfactory as a single detector system. The two detectors may have slightly different response characteristics, which will mean that the DC offset signal is not exactly cancelled out.

The stability of a single column system has been considerably improved by the introduction of Mass-Flow Regulators which control the *mass* of gas flowing instead of just the pressure of the supply or the volume flow-rate. This reduces any DC drift or offset in the gas-flow rate, and hence it is often possible to use a single-column (single-beam!) system.

5.2.4. Use of Microprocessor Memory

We have already discussed how a differential method can be used to eliminate DC offset, which may occur in an instrument as a result of changes of the experimental parameters.

For the gas chromatography system, the changing characteristics of the sample column as a function of temperature can be balanced by using a 'dummy' column. A single-beam spectrophotometer which is used by scanning different wavelengths must be re-calibrated at each new wavelength because of the variations of source intensity and detector sensitivity. This is avoided in a double-beam system by the use of a second or reference-beam.

If the differential method is *not* used, the output from the detector will have an offset which changes as the instrument sweeps through its optical range – see Fig. 5.2e. This is often referred to as *Baseline Drift*.

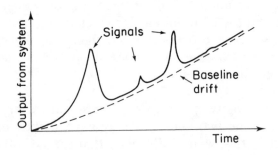

Fig. 5.2e.   *Baseline drift*

It is important to distinguish between two very different situations concerning drift:

(*a*) the drift is due to the random fluctuations of $1/f$ noise (see Section 3.5.3),

(*b*) the drift is directly related to the position on the sweep range, within the instrument.

In (*a*) the drift obviously varies in a way that cannot be predicted, and its effect can be overcome only by using the types of techniques described earlier, for example, modulation, differential methods, signal averaging. In (*b*) the drift will be *reproducible* every time the sweep is repeated, provided, of course, that the experimental conditions and the properties of the relevant instrument components remain constant.

It is in this latter situation that a microprocessor memory system can be used. Initially, the instrument performs a measurement sweep in the absence of the sample. This type of measurement would be expected to result in a plot of the DC Offset or Baseline of the output as shown in Fig. 5.2e. However, instead of this being plotted out onto a recorder, the output voltage of the instrument is recorded at several hundred points across the range, and the value of each of these voltages is fed into the memory of a microprocessor. The instrument is then used to repeat the experiment, this time with the sample, producing a completely new set of output voltages (one for each of the record-points). The difference between the two voltages (baseline and baseline plus sample) gives the output signal due to the *sample*. The instrument with this type of memory system may either perform the baseline correction automatically as soon as it is switched on, or it may be necessary for the operator to 'instruct' the instrument to carry out the process and store the information.

It is because of the increased *stability* of many of the important components in instruments, together with this new ability of computer systems to remember reproducible drift variations, that many new instruments can now be constructed by using the *single-beam* system.

5.2.5. Conclusion

We have now introduced three very different types of analytical instrument. The pH-meter relies on the electrode for the initial step in the measurement of the unknown pH, and then the meter itself is entirely composed of electronic systems. The spectrophotometer, however, is a very complex instrument with optical, mechanical, and electronic components in a closely inter-linked system. The degree of sophistication is such that many modern systems now function almost automatically under the control of a microprocessor. The gas chromatograph is a different type of instrument again, because of the wide range of decisions (concerning materials, etc) that must be made by the operator, and the variety of functions that must be controlled. We have also examined briefly the way in which modern 'microchip' technology can radically reform the actual design of the instruments themselves.

SAQ 5.2a   Indicate which of the following statements about
the slope-control of the pH meter are true.

If the slope-control is changed from 100% to
80%, the effect is:

(*i*)   to decrease the input impedance of the instrument.

                                                T / F

(*ii*)   to increase the input impedance of the instrument.

                                                T / F

(*iii*) to decrease the gain of the amplifier.

                                                T / F

(*iv*) to increase the gain of the amplifier.

                                                T / F

SAQ 5.2b

A computer-memory system as described in Section 5.2.4 is used to reduce the effect of various types of drift in instruments. A list of several forms of drift is given below. Indicate by ticking the 'Y', the types of drift whose effect can be reduced in this way.

Spectrophotometer.

(*i*) Variation of light-source intensity with mains voltage.

Y / N

(*ii*) Variation of detector sensitivity with wavelength.

Y / N

(*iii*) Variation of amplifier-gain with time.

Y / N

Gas Chromatograph.

(*iv*) Variation of detector sensitivity with flow-rate,

Y / N

(*v*) Variation of the physical properties of the column with temperature,

Y / N

(*vi*) Variation of the flow-rate controller with time.

Y / N

5.3.   SPECIFICATIONS OF AN INSTRUMENT

5.3.1.   Introduction

The *specifications* of an instrument are intended to tell us 'what the machine will do'. This will include many factors, eg:

the various types and ranges of measurement possible, ie the performance characteristics,

the quality of performance (accuracy, linearity, etc),

technical details such as electric power supply requirements.

It is not possible in this section to cover all the different items that may be given in the specifications of an instrument. We shall, instead, select some important examples that relate to the topics that have been introduced in the earlier parts of the unit.

You will notice that in this section we do not always use a consistent terminology. For example, 'detectability' and 'minimum detection limit' are both used to describe the same characteristic. This variation reflects the lack of uniformity that actually exists in manufacturers' specifications. It is important that you learn how to check on the true meaning of an expression by referring to the units used.

5.3.2.   Performance Characteristics

We can think of an instrument, under the control of an operator, as a total system which expresses the measured analytical quantity, Q, as an output signal, S. The output signal, S, is a function, $F(Q)$, of the analytical quantity, Q. We saw in Section 1.3.4 that the variables, S and Q, may be in very different forms, eg the deflection (S) of a pen on a recorder may represent absorbance (Q). However, in many instruments, the output signal, S, is presented in such a way that it is in a form which equates *numerically* to the value of Q. This is achieved either electronically, by presenting the result in a digital read-out, or else the result is presented against a scale (eg wavelength or pH) which is calibrated directly in the value of Q.

An instrument (or separate functional element) is said to have a *linear* response if the output, S, is directly proportional to the input, Q. The output and input can then be related by the familiar 'straight-line' equation below.

$$S \ = \ mQ \ + \ C$$

'C' is a Zero Offset (see Section 3.2.3) which is usually set to zero.

If S and Q are expressed in different forms, the *sensitivity* of the instrument is given by

Sensitivity $= \ S/Q \ = \ m$, with m in appropriate units.

For example, if S is a displacement (cm) on a chart recorder, and Q is the quantity of an unknown analyte expressed in μg, then the sensitivity, m, is expressed in units of cm μg^{-1}. See also Section 1.3.4. Sometimes the 'sensitivity' of an instrument is expressed as the reciprocal of this definition of sensitivity. Nevertheless, it is always possible to check which one is being used by looking at the units of the expression.

∏   Two instruments are designed to measure the concentration of a particular element, and they present the result as a displacement of a pen on a chart recorder. In their respective specifications, the 'sensitivities' are quoted as below:

Instrument A – 30 μg cm^{-1}

Instrument B – 50 cm mg^{-1}

Which of the two is the most sensitive?   A / B

By looking at the units used, we can see that instrument B has been specified by using 'sensitivity' as described in the text, ie the output signal divided by the input signal. However, the 'sensitivity' for A has been defined by the reciprocal of this expression. To compare the two instruments we must convert the specifications into the same format.

The 'true' sensitivity for A $= 1/30$ cm μg^{-1}

$$= 1/0.03 \text{ cm mg}^{-1}$$

$$= 33 \text{ cm mg}^{-1}$$

Now that the values for the two instruments are in the same units, an immediate comparison reveals that 'B' has the greater sensitivity according to the definition in the text. Instrument B is the more sensitive.

If S is calibrated directly in terms of Q then the reading given by S numerically equals the value of Q. In this situation, the value of m is equal to one. It is not possible then to describe the 'sensitivity' of the instrument as the value of m, and, instead, it is common to express the 'sensitivity' as the minimum value of Q that can be detected by the instrument – the *Detection Limit*, $Q(\text{min})$.

∏   Calculate the sensitivity of a Flame Ionisation Detector (FID) in Cg^{-1} (C $=$ coulombs) of carbon, if the Detection Limit is 2.0×10^{-13} gs^{-1}, and this corresponds to an output current from the detector of 3.0×10^{-15} A.

We know from the question that an amount of carbon equivalent to 2.0×10^{-13} gs^{-1} will give a current of 3.0×10^{-15} A. This current is equivalent to a flow of 3.0×10^{-15} coulombs in each second (see Section 2.1.2).

The current of 3.0×10^{-15} Cs^{-1} is the Output Signal. The mass flow of 2.0×10^{-13} gs^{-1} of carbon is the Input Signal.

Thus,

Sensitivity $=$ Output Signal/Input Signal

$$= 3.0 \times 10^{-15} \text{ } Cs^{-1}/2.0 \times 10^{-13} \text{ } gs^{-1}$$

$$= 0.015 \text{ } Cg^{-1}.$$

An important measure of the versatility of an analytical instrument is given by the *Dynamic Range*. It is obviously useful to have an instrument which is able to measure a concentration as low as (for example) 1 ppm, but it would then be a great limitation on the use of that instrument, if a sample with a concentration greater than 10 ppm was *too* concentrated to be measured directly. If the minimum detectable signal is $Q(\text{min})$ and the maximum measurable signal is $Q(\text{max})$ then the *dynamic range* of the instrument is defined as follows.

$$\text{Dynamic Range} = Q(\text{max})/Q(\text{min})$$

An instrument in which the dynamic range is large is more versatile than an equivalent instrument in which it is small.

∏   An instrument which measures the quantity of material, X, has a chart recorder output with a full width of 20 cm, and a sensitivity ranges – 10, 20, 50, and 100 cms μg^{-1}.

The noise on the chart recorder output has an amplitude of 3.0 mm when the instrument is set to its most sensitive range.

(*i*) Calculate the Detection Limit corresponding to a signal-to-noise ratio of 2 : 1.

(*ii*) Calculate the Dynamic Range of the instrument by using the Detection Limit calculated in (*i*).

(*i*) The sensitivity ranges are given in units of cm/μg^{-1}, with the most sensitive range having a displacement of 100 cm for a quantity of X equal to 1 μg.

Thus,

Maximum sensitivity $= 100$ cm μg^{-1}

$$= 1000 \text{ mm } \mu g^{-1}.$$

A signal-to-noise ratio of $2:1$ means that the signal amplitude is twice the noise amplitude. The noise amplitude is 3.0 mm, thus the amplitude of the Signal corresponding to the Minimum Detection Limit is 2×3.0 mm $= 6.0$ mm. This minimum signal is observed on the most sensitive range.

We know from above that 1000 mm is equivalent to 1 μg.

$\therefore$ 6.0 mm is equivalent to $6/1000 = 6.0 \times 10^{-3}$ μg

Thus the Minimum Detection Limit, $Q(\text{min}) = 6.0 \times 10^{-3}$ μg.

(ii) On the least sensitive range (10 cm μg^{-1}), the maximum acceptable analytical input $[Q(\text{max})]$ will cause a full-scale deflection (20 cm).

Thus,

$$Q(\text{max}) = 20/10 \ \mu g = 2.0 \mu g$$

The Dynamic Range is given by,

$$Q(\text{max})/Q(\text{min}) = 2.0/6.0 \times 10^{-3} \simeq 3.3 \times 10^{-2}$$

(within the accuracy of the quoted figures).

5.3.3. Resolution and Selectivity

Resolution and selectivity both imply an ability of the instrument to differentiate between different constituents in the same sample. However, 'selectivity' is usually used if there are some constituents to which the instrument is designed *not* to respond, whereas 'resolution' is used if the instrument can, by altering the operational conditions, respond independently to different constituents. This subtle difference is explained further by examples below.

We met, in Section 1.3.1, for example, ion-*selective* electrodes which are sensitive to sodium but not to potassium ions. If $Q(\text{Na})$ and $Q(\text{K})$ are respectively the concentration of sodium and of potassium ions in a sample, the output of a sodium ion-selective electrode would be given by,

$$S = m(\text{Na})\, Q(\text{Na}) + m(\text{K})\, Q(\text{K})$$

where $m(\text{Na})$ and $m(\text{K})$ are the sensitivities of the electrode to sodium and potassium ions respectively. We normally expect that, for the sodium-selective electrode,

$$m(\text{Na}) \gg m(\text{K})$$

The selectivity of the electrode is often expressed as the ratio of the two sensitivities:

$$\text{Selectivity} = m(\text{Na})/m(\text{K}).$$

We normally use the term 'selectivity' if the actual selection ability is a fixed function of the components of the instrument, so that we can change from one selectivity to another only by *physically replacing* one part of the instrument. In gas chromatography, a Nitrogen-Phosphorus Detector (NPD) is used to respond selectively to nitrogen and phosphorus, with only a very low response to other constituents. If we wish to measure a constituent other than nitrogen or phosphorus, we must physically replace the NPD with another detector more responsive to the required constituent.

∏   The selectivity of a NPD is given as N : Hydrocarbon = $10^4 : 1$.

    If the Minimum Detectability of nitrogen is 8×10^{-13} g nitrogen s^{-1}, estimate the Minimum Detectability of Hydrocarbons.

The expression for the selectivity tells us that the sensitivities of the detector to nitrogen and hydrocarbon respectively are in the ratio of $10^4 : 1$. Thus the answer is going to be one of the values 8×10^{-17} or 8×10^{-9}.

We know that the NPD is designed to detect nitrogen, hence the sensitivity to hydrocarbons will be low. With a lower sensitivity the minimum detection limit for hydrocarbons will be greater than for nitrogen. Hence 8×10^{-9} is the correct answer.

The 'resolution' of an instrument implies the ability of the instrument to discriminate between two quantities by *adjusting* the experimental conditions within the instrument, and *without* exchanging any components. In a gas chromatograph, when we have selected a relevant detector, we still have the ability to adjust other experimental parameters (eg gas-flow and temperature programming) to alter the times at which *different* constituents appear at the detector. These factors affect the resolution of the instrument.

We now continue to discuss ideas of resolution as applied to spectroscopy, where the adjustable variable is usually the wavelength of the radiation. A signal output from an instrument may be presented as a voltage, or the displacement of a recorder pen, which changes as a function of some system variable such as time or wavelength. A typical example is given in Fig. 5.3a, which shows a signal, or 'line', produced by a spectroscopic instrument and drawn out as a function of wavelength. In spectroscopy, each separate signal output is frequently called a spectrum 'line' because of the equivalent lines that were produced on the photographic plates of early optical spectrophotometers.

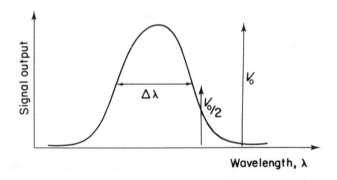

Fig. 5.3a.   *Spectral line output from spectrophotometer*

Every line signal must have some width, and we have already seen that it is convenient to use the idea of 'width at half height', $\Delta\lambda$, as shown in the diagram. The peak height, V_0, of the line is also shown.

It is not always easy to identify the presence of a signal. In Section

4.2.3, we introduced the idea that sometimes it is difficult to tell the difference between an analytical signal and a noise signal. We now introduce the idea that sometimes it is difficult to tell the difference between one analytical signal and another.

Π   How many separate 'lines' are visible in the spectrum given in Fig. 5.3b?

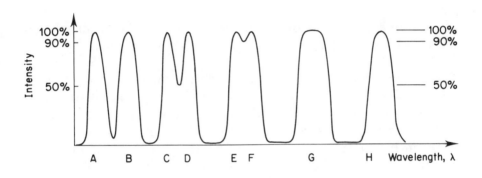

Fig. 5.3b

It is very difficult to be exact about the answer to this question.

Clearly A and B are two separate lines.

It is also easy to see that C and D are lines which have overlapping tails.

Similarly you could say with some confidence that E and F were also separate lines but with a considerable amount of overlap.

But what of the line G? Is this two lines very close together, or is it just a single line with an unusual shape?

Is H just the single line that it appears to be, or is it two lines very close together indeed?

Depending on how you answered the question about G and H you may have counted 8, 9 or even 10 lines.

The previous question illustrated some of the problems concerned with being able to identify two lines as being separate. These are the problems of *resolution*.

In the question, the lines A and B are completely *resolved*, ie their existance as two separate lines is quite obvious. Lines C and D are also said to be resolved. The 'dip' in intensity between the two lines is 50% of the height of the 'peaks'. Lines E and F are only just resolved. The 'dip' in this case is only 10% of the peak-height.

In G there is no visible 'dip' in the top of the combined signal, and we would say that the two lines have *not* been resolved.

There is no *absolute* criterion by which one may judge whether or not two lines have been resolved. A minimum criterion would be that there should be an identifiable 'dip' between the two peaks, and, in practice, this dip must be at least 10% of the peak height as in E and F. In optics, the Rayleigh Criterion for resolution requires that this dip should be about 20% of the peak height.

However, a more workable, day-to-day, criterion would be that the dip must be at least 50% of the peak height if two adjacent lines are to be resolved satisfactorily. If the 50% criterion is used then the lines C and D are only just resolved, whereas we would say that the lines E and F are unresolved. In fact, if you used this criterion for the question, you would have recorded only 7 lines!

∏   A monochromator is an optical device which allows only a narrow band of wavelengths to pass and blocks all other wavelengths. It behaves in a way similar to a tuned filter (see Section 3.3.3) and can be characterised as having a band-width, $\Delta\lambda$, and operating at a wavelength λ_o, see Fig. 5.3c. The diagram shows the output from a monochromator (of band-width 0.4 nm) which has been illuminated with a source of white light, and has equal intensity at all wavelengths.

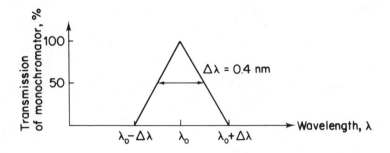

Fig. 5.3c. *Transmission of monochromator*

What will be the output if *monochromatic* radiation of wavelength 500 nm is used as the source, and the monochromator is continuously adjusted so that its operating wavelength, λ, is swept across the range from 498 nm to 502 nm.

Assume that the band-width of the monochromatic radiation is very much less than 0.4 nm.

Choose your answer from (i), (ii), (iii) or (iv) in Fig. 5.3d.

We can see from Fig. 5.3c that if the monochromator is set at a wavelength, λ_0 it will not pass any radiation whose wavelength, λ_s, is more than $\lambda_0 + \Delta\lambda$ or less than $\lambda_0 - \Delta\lambda$.

Thus if the monochromator is set at 499.5 nm, it will *not* pass any of the source radiation ($\lambda_s = 500.0$ nm) because,

$$500.0 \text{ nm} > (499.5 + 0.4) \text{ nm}$$

However, in each of the graphs, (i), (ii) and (iii), there *is* a non-zero intensity of radiation at 499.5 nm.

Thus (i), (ii) and (iii) are all wrong.

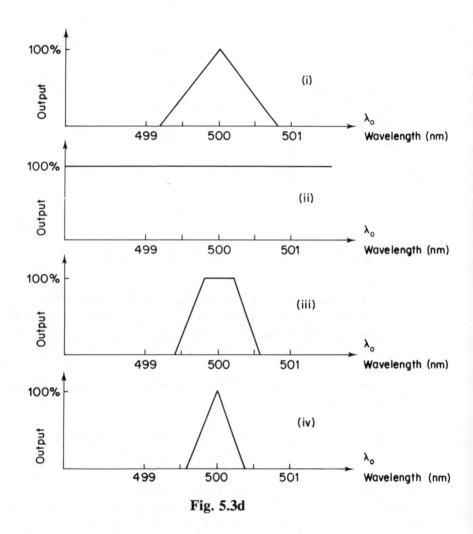

Fig. 5.3d

As the monochromator wavelength, λ_o, is changed it will start to transmit radiation of wavelength, $\lambda_s = 500.0$ nm, when the control setting is at $500.0 - 0.4 = 499.6$ nm.

The intensity transmitted will increase linearly (as in Fig. 5.3d(iv)) as a function of the difference between the source wavelength, λ_s, and the control setting of the monochromator, λ_o. It will reach a maximum when λ_o and λ_s are the same, and then beyond this point the transmission will decrease.. The correct answer is (iv).

∏   The same monochromator as used in the previous question is illuminated with a sodium-light source which has two monochromatic lines at 589.0 nm and 589.6 nm respectively. The setting of the monochromator is made to sweep from 588 nm to 591.0 nm.

What will be the output? Choose your answer from (*i*), (*ii*), (*iii*) or (*iv*) in Fig. 5.3e.

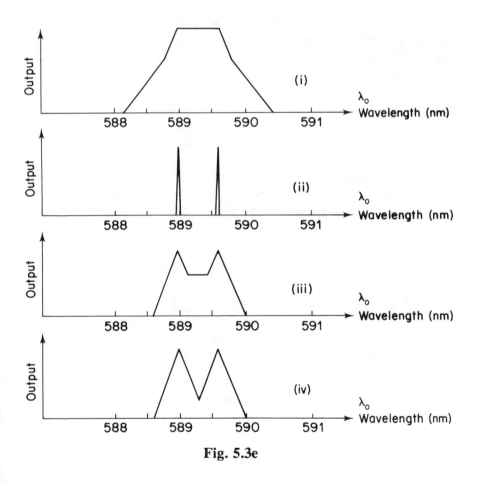

Fig. 5.3e

Initially we can treat each line, λ_{s1} (= 589.0 nm) and λ_{s2} (= 589.6 nm), independently as in the previous question. They each produce an output with a band-width of 0.4 nm.

Hence, choice (*ii*) is incorrect because the band-widths of the two 'lines' are too small.

The true output is shown in Fig. 5.3f.

It can be seen that there is an overlap of the two lines, and in the region of overlapping the two intensities must add together. The correct answer is (*iii*).

Choice (*iv*) is incorrect because no addition has been made in the region of overlap. Choice (*i*) corresponds to a band-width for each 'line' of 0.8 nm, and would result from a monochromator whose band-width is 0.8 nm.

By comparing the two outputs in (*i*) and (*iii*), you can see that for a band-width of 0.8 nm, the two lines are not resolved, but for a band-width of 0.4 nm, the two lines are resolved.

You can probably infer from this that, in order to resolve two lines separated by $\Delta\lambda$ nm, the maximum permissible band-width for the monochromator must be also $\Delta\lambda$ nm.

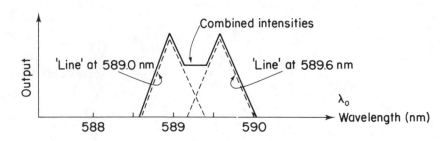

Fig. 5.3f

5.3.4. Noise and Drift

We discovered earlier in this unit (Section 3.5) that the term 'noise' covers *all* unwanted signals, including 'drift' (Section 3.2). However, it is common, in the specifications for instruments, to use 'noise' as a name which covers random fluctuations at frequencies greater than those of the analytical signals (typically greater than 0.1 to 1 Hz). 'Drift' is then used to describe the very low frequency fluctuations which occur over periods of seconds up to hours and days.

It is almost impossible for a manufacturer to give a specification for noise and drift which covers all the operational conditions of the instrument. The noise level at the output of an instrument may be changed drastically by the setting of the controls, and also the level of the analytical signal. For example, the noise may increase considerably on the most sensitive ranges, at the extremes of a wavelength sweep, or for very absorbent samples in an optical-null spectrophotometer (see Section 5.1.3).

∏ A particular single beam spectrophotometer, with an output calibrated from 0 to 100% transmittance, suffers from both random noise and drift. The magnitude of the random noise signal is constant and is equivalent to about ± 0.1 at the output. The error in $\%T$ due to noise is thus,

$$\Delta T(n) = \pm 0.1$$

The drift (about 1%) occurs in the intensity of the lamp source, and thus gives an error which is proportional to the output signal from the detection system. The error in T thus increases with the magnitude of T,

$$\Delta T(d) = \pm 1/100 \times T = \pm 0.01 \times T$$

Calculate these two errors, and the corresponding fractional errors, $\Delta T/T$, for different values of the transmittance T, and complete the table below.

The significance of '3.16' and '31.6' will appear in a later question.

T%	$\Delta T(n)$	$\Delta T(n)/T$	$\Delta T(d)$	$\Delta T(d)/T$
1	...	...	0.01	0.01
3.16	...	...	...	...
10	0.10	0.01	...	...
31.6	...	...	...	...
100	0.10	...	1.0	...

The correct values for the errors and fractional errors are given below:

T%	$\Delta T(n)$	$\Delta T(n)/T$	$\Delta T(d)$	$\Delta T(d)/T$
1	0.10	0.10	0.01	0.01
3.16	0.10	0.0316	0.0316	0.01
10	0.10	0.01	0.10	0.01
31.6	0.10	0.00316	0.316	0.01
100	0.10	0.001	1.0	0.01

Note that, in this example, the fractional error due to the 'noise' decreases for larger values of T, and that the fractional error due to 'drift' is constant.

Note that if we multiply the fractional error by 100 we get the *percentage* error, $(\Delta T/T) \times 100\%$.

We met in the question in Section 2.4.3 the concept of Absorbance, A, which is related to Transmittance by the equation::

$$A = \log(100/T)$$
$$= -\log(T/100).$$

The value of absorbance is more directly useful to the analyst than the transmittance because, in simple situations, absorbance is proportional to the concentration of a solution.

To find the effect of errors in the instrument we can differentiate the above expression, and arrive at the equation

$$\frac{\Delta A}{A} = \frac{-1}{2.3\,A} \times \frac{\Delta T}{T}$$

Thus the fractional (or percentage) error in A =

$$\frac{-1}{2.3\,A} \times \text{Fractional (or percentage error) in } T.$$

We have left the minus sign in the equation, but, since it is impossible to predict the direction of errors, the actual sign is irrelevant.

∏   Take the results from the previous question for the noise and drift errors, and calculate the respective percentage errors in A for different values of A. Plot your results on the graph provided in Fig. 5.3g.

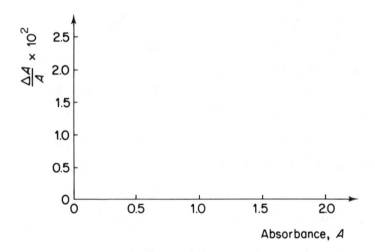

Fig. 5.3g

Note:

$T = 31.6\%$ corresponds to a value of $A = 0.5$

$T = 3.16\%$ corresponds to a value of $A = 1.5$

The correct graph is shown in Fig. 5.3h.

You can see that the different types of errors have different magnitudes depending on the experimental conditions, viz. the value of the absorbance of the sample A. The actual error in the instrument is the *sum* of all errors.

It can be seen from Fig. 5.3h, that, for this particular instrument, the value of A which has the minimum fractional error will be around 1.0. You must bear in mind the optimum accuracy range for your own instrument when you are preparing your sample for measurement.

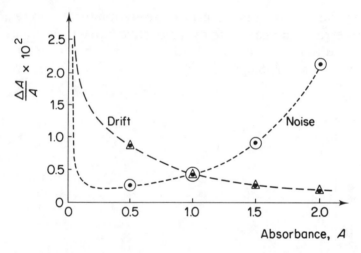

Fig. 5.3h.   *Fractional errors due to noise and drift*

5.3.5.   Instrumental Errors

The *total error* in an analysis is the difference between the value of the relevant quantity (eg the concentration of an element in an unknown sample) obtained by the analytical procedure and its *true* value. The *accuracy* of the analytical procedure is a measure of the *maximum* total error that can be expected to occur, and thus it is the ultimate assessment of the quality of an analysis.

The errors that arise in an overall analytical procedure originate from many factors including sampling, sample preparation, calibration and interpretation. These will be discussed extensively in other units. The errors that occur within the instrument itself are only a part of the total error in the analysis.

In their advertising literature, it is common for the manufacturers of an instrument to describe the maximum total error of the instrument as its 'accuracy'. This may sound more flattering to the instrument than using the term 'error', however, you must remember that 'accuracy' actually represents an *error* that the instrument may *add* to the other errors in the overall analysis.

∏   You have just bought a new instrument and you wish to know its accuracy in performing a particular type of measurement.

Can you obtain this information:

(*i*) by making up a known standard sample, performing a measurement, and calculating the 'accuracy' as the difference between the known and the measured value, or,

(*ii*) by consulting the manufacturer's specifications for the 'accuracy' of the instrument?

You will find that some books define accuracy as being the difference between a 'measured' and a 'true' value. However, it is possible that, *by chance*, your measured value in (*i*) is very close indeed to the true value, but this does not mean that your instrument is accurate. Performing the same measurement at another time, you may find a very different value.

A single measurement, as in the question, is not enough. If you were to *repeat* this measurement many times you could then begin to get an idea of both the random errors and *total error* ('accuracy') of the instrument as shown in Fig. 5.3i. Indeed, to obtain a reasonable idea of the accuracy of the instrument, you must repeat the *whole* procedure regularly so that you monitor any drift that may occur in the instrument.

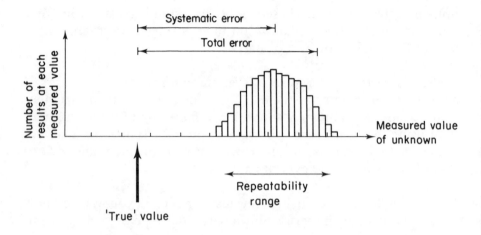

Fig. 5.3i.   *Random and systematic errors*

The manufacturer's specifications in (*ii*) *will* give some information about the minimum level of accuracy. However, as we have seen, the accuracy of an instrument changes under different conditions and it is advisable to consider an objective testing procedure similar to that discussed above.

There is a very real danger that inexperienced operators of instruments assume that the 'accuracy' of the overall *analysis* can be equated to the quoted accuracy of the *instrument*. This is very far from the truth, and you should continue to interpret instrumental 'accuracy' as being the error *contribution* made by the instrument to the overall error in the analysis.

The total instrument error arises as the sum of separate errors occurring in different aspects of the instrumental measurement. We discuss below some factors which contribute to the maximum (total) error and which may be also described in the specifications of the instrument.

If the same measurement is repeated several times, we would expect the final result to be slightly different each time because of varying, or random, errors (eg noise). The *spread* of these different results is the *repeatability* of the measurement. Repeatability gives a good idea of the *random* errors inherent in the analysis. However, it tells us nothing about any *systematic* errors that may be present. These are errors which are the *same* every time the experiment is repeated (see Fig. 5.3i).

You may often find it quite possible to *read* the value of the relevant output signal to an *apparent* 'accuracy' which is far greater than the real accuracy of the instrument. It is, for example, possible to read the scale of a meter to about one half of the smallest division, and this is independent of the accuracy of the instrument. Thus a scale, 0 to 100 units, with a division spacing of 1 unit, will have a *readability* of ±0.5 units (see Fig. 5.3j).

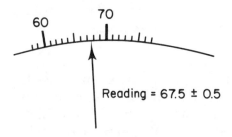

Fig. 5.3j.   *Readability of scale*

Readability is a measure of how precisely you can record the output of the instrument. Note that accuracy cannot be better than readability, but readability can be better than accuracy.

∏   An atomic absorption spectrophotometer uses a 4 digit digital display read-out to present the relative absorbance of the sample.

    A standard sample with a constant absorbance is measured several times, giving the following results for the absorbance.

    0.522A, 0.524A, 0.527A, 0.521A, 0.517A, 0.519A, 0.521A, 0.523A.

From these results give an *estimate* for the,

(*i*)   the readability,

(*ii*)   the repeatability,

of the instrument.

What can you say about the *accuracy* of;

(*iii*)   the instrument

(*iv*)   the value that could be derived from these for the concentration of the sample.

The answer to (*i*) is relatively simple. It is easy to read a difference of '1' in the last digit of the display, eg to differentiate between 0.521A and 0.522A. Hence the *limit* of readability is equivalent to a possible error of ± 0.001A. You may have decided that the possible error in reading is ± 0.0005A. This would in fact be true if we could be sure that the electronics of the system 'round-up' and 'round-down' any figures in the fourth decimal place, so that the display gives the closest value in the third (and last visible) decimal place. We cannot be sure that this is true, so it is best to be safe and say that the possible error in this display is ± 1 in the last digit. Thus the readability is ± 0.001A.

The repeatability is given by the spread of *possible* results. Of course, there is only a limited number of results given here, and so we can only make a rough estimate of repeatability. Assuming that no operational or human errors have crept into the results, the spread of instrumental results is from 0.517A to 0.527A and is equivalent to ± 0.005A. Thus we can estimate that the repeatability is *not less* than ± 0.005A.

The only comment that we can make about the accuracy of the instrument, (*iii*), is that the total error must be greater than ± 0.006A. This figure is due to the contribution to total error by the repeatability and readability. There will also be systematic errors in the instrument which must be included, and which will further increase the total error, and limit the accuracy of the measurement.

The actual value of the concentration of the sample, (*iv*), must be uncertain by a possible error which is greater than the error in (*iii*). There will also have been errors in the assumptions and techniques used to derive the value of concentration from the absorbance values given by the instrument. For atomic absorption instruments, the reduction of these errors is particularly dependent on the skill and experience of the operator of the instrument.

One possible error that can occur in an instrumental system which is supposed to have a linear relationship is that the response of the instrument is actually *non-linear*. This situation is shown in Fig. 5.3k.

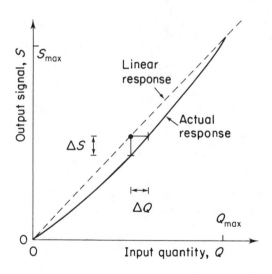

Fig. 5.3k. *Non-linearity of response*

The *linearity* of the system is normally defined as follows.

Linearity = Maximum error, ΔQ/Measurement range, $Q(\text{max})$

Alternatively, the linearity could be defined as follows,

Linearity = Maximum error, ΔS/Output range, S(max)

giving the same value as the definition in terms of ΔQ and Q.

If the non-linearity is small then we can normally approximate the curve as a simple quadratic function. In other words, the curve is a segment of a very large circle (a situation which is exaggerated in Fig. 5.31).

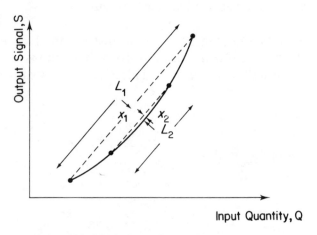

Fig. 5.31.   *Reduction of linearity errors*

If L is the length of a segment, and the deviation from a straight line is x, then geometry tells us that for $x \ll L$,

$$x \propto L^2$$

Thus if we reduce the segment length by one half, we reduce x by one quarter.

The error due to non-linearity can, for small deviations, be reduced in the same ratio as the *square* of the range between calibration points:

$$x_2/x_1 = (L_2/L_1)^2$$

Thus it is desirable to choose calibration points which are as close together as possible.

∏   The light detection system of a simple colorimeter has a non-linear response.

The instrument is calibrated at 100% T by using a reference 'blank'.

(*i*)   If the linearity is quoted as being 1%, calculate the resultant *percentage* error in the measured transmittance ($\%T$) of a sample whose transmittance is approximately 50%.

(*ii*)   How could the measurement method be improved to reduce the errors arising from non-linearity.

The measurement range in (*i*) is 100 (in $\%T$), and, by using the definition of linearity given above, we can calculate that the maximum error in T will be ±1.0.

The maximum error occurs in approximately mid-range – ie at 50% T. Thus the *percentage* error may be as great as

$$1/50 \times 100 = 2\%.$$

Note that the actual percentage error (2%) in a particular result may be greater than the error (1%) given for non-linearity in the system.

For the answer to (*ii*), we saw in the text that the error due to non-linearity can be drastically reduced by bringing the calibration points closer together.

We can illustrate the answer by considering calibrating the colorimeter at 100% T with the blank, and at 50% T with a standard solution. The calibration at 0% T should also be checked by completely blocking the light to simulate zero transmission. We do not now expect to have any error due to non-linearity when we measure an unknown sample near 50% T, because this is one of our calibration points. (By the way, this is very similar to the situation in which we replace a standard with an unknown sample, ie it is equivalent to a substitution measurement method, see Section 2.4.4.)

If we now assume the output of the colorimeter to be linear between calibration points, the maximum errors due to non-linearity probably occur at about 25% T and 75% T, Fig. 5.3m,

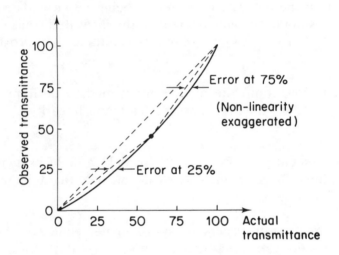

Fig. 5.3m.   *Calibration at 0%, 100% and about 50% T*

The range between calibration points is now 50. Thus we have reduced the range (from 100 to 50) by a factor of 2. Using the theory in the text, the error due to non-linearity will be reduced by $2^2 =$ 4. Hence the possible error for the colorimeter in the question, at 25% T and 75% T, is $\pm$ 1.0/4 = $\pm$0.25.

If we convert this into percentages,

$$\text{Error at } 25\% T \;=\; \frac{0.25}{25} \times 100 \;=\; 1\% \text{ in } T.$$

$$\text{Error at } 75\% T \;=\; \frac{0.25}{75} \times 100 \;=\; 0.33\% \text{ in } T.$$

5.3.6. Conclusion

Both in this section, and throughout the unit, we have introduced a range of different types of analytical instruments. We have also tried to emphasise the very important role that the operator plays in the efficient use of the instrument, and as early as Section 1.4.4 we introduced the idea that the operator was a component part of the instrumental process. The skill and experience of the operator will markedly affect the quality of the analytical process.

After having examined the specifications of any instrument, you should also examine your own 'specifications'.

Do you understand the operation of the instrument sufficiently well to operate it efficiently and accuracy?

Are you able to spot instrumental errors developing in the analysis?

This unit has tried to give a grounding of some of the important topics in chemical instrumentation. You should now study, very carefully, the separate section on the instruments that occur in the units devoted to particular techniques.

SAQ 5.3a	The Specifications for a particular Y-t Chart Recorder contain the following entries.

Chart Width	200 mm
Writing Speed	400 mm . s^{-1}
Ranges	20, 50, 100, 200 μV . mm^{-1}
	0.5, 1, 2, 5, 10
	20, 50, 100, 200
	500 mV . mm^{-1}

$\longrightarrow$

SAQ 5.3a
(cont.)

Accuracy at FSD	$\pm 0.2\ \%$
Linearity	0.1 % FSD
Repeatability	0.1 % FSD
Input Impedance	1 M ohm on all ranges
Chart Drive Speeds	2, 5, 10, 20 mm . min^{-1}
	1, 2, 5, 10, 20, mm . s^{-1}
Zero Setting	Adjustable within full scale.

(i) Estimate the frequency band-width (maximum frequency response) for a varying signal which has an almost full scale deflection.

(ii) Calculate a value for the maximum sensitivity of the instrument.

(iii) What is the possible error (in mV) due to non-linearity on the 1 mV . mm^{-1} range?

(iv) If the minimum detectable signal is equivalent to 0.2 % FSD, calculate the dynamic range of the instrument.

(v) A signal is drawn out with a line-width of 16 mm when the speed of the chart paper is 10 mm . s^{-1}. Calculate the band-width of the signal in the frequency domain.

$\longrightarrow$

SAQ 5.3a
(cont.)

   (*vi*)   Estimate the possible error, arising from relative input and output impedances, if the recorder is used to measure a voltage originating from a source with an output impedance of 10 kΩ.

  (*vii*) What is the maximum output impedance of a measured source voltage if the resultant error, as in (*vi*), is to be less than the other total errors in the recorder.

SAQ 5.3b   The Specifications for an Infra-red Spectrophotometer contain the following entries.

Principle	Double-beam optical-null
Digital Display	4 digit LED display of wavenumber
Abscissa Ranges	$4000–2000$ cm^{-1},
	$2000–200$ cm^{-1},
	running consecutively.
Abscissa Accuracy	$4000–2000$, $\pm$ 6 cm^{-1}
	$2000–200$, $\pm$ 3 cm^{-1}
Abscissa Repeatability	Within 0.6 mm of chart
S/N	Better than $100:1$, outside the region of atmospheric absorption and with narrow slit programme.
Resolution	Narrow Slit Programme 3 cm^{-1} at 1000 cm^{-1} 5 cm^{-1} at 3000 cm^{-1}
Chart Size	15 cm $\times$ 47 cm (see Fig. 5.3n.)

$\longrightarrow$

SAQ 5.3b (cont.)

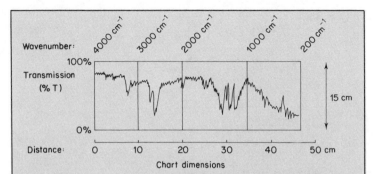

Fig. 5.3n

Note. In Infra-red spectroscopy, the 'wavelength' scale is not calibrated in nanometres. The scale is actually *inversely* proportional to the wavelength, and is calibrated in 'wavenumbers' which have units of cm^{-1} and are obtained by taking the reciprocal of the wavelength expressed in centimetres;

$$\text{Wavenumber } (cm^{-1}) = 1/\text{Wavelength (cm)}$$

It is also common, for technical reasons, for the wavenumber scale in an ir spectrophotometer to be divided into two ranges.

(*i*) What is the wavenumber readability in cm^{-1} on the digital display?

(*ii*) Calculate the repeatability in cm^{-1} for the two ranges on the chart.

(*iii*) In which range does the instrument have the better resolution?

(*iv*) Why does the signal-to-noise ratio decrease if there is absorption of radiation in the atmosphere (which would affect both beams similarly), or if a narrow-slit programme is used in the monochromator?

SAQ 5.3b

SAQ 5.3c   The Specifications of a pH meter contain the following entries.

Read-out	$3\frac{1}{2}$ digit display
Ranges	0 to 14.00 pH
	0 to 1900 mV
	−30 to 105 °C
Automatic Temperature Compensation	0 to 100 °C
Manual Temperature Compensation	0 to 100 °C
Slope Correction	80 to 110 % $\longrightarrow$

**SAQ 5.3c
(cont.)**

Accuracy	± 0.06 pH
	± 6 mV
Repeatability	± 0.02 pH
	± 2 mV
Linearity	± 0.04 pH
	± 4 mV
Input impedance	$> 10^{13}$ ohms
Recorder Output	$10 \text{ mV} . \text{pH}^{-1}$
Response Time	10 ms

Note that the pH-meter can be used either to record pH directly, or to record the voltage (in mVs) from other types of electrodes. It can also be used to measure temperature.

(i) If the input impedance could be increased by a factor of 10 from 10^{13} to 10^{14} ohms, by what factor would the accuracy of the instrument change?

(ii) If the instrument is calibrated at pH 4 and pH 9, estimate the error due to non-linearity in a reading at a value of about pH 6.5.

(iii) Is it essential to use an automatic temperature probe with this instrument?

(iv) If we wish to display the full pH range (0 to 14) on the recorder described in SAQ 5.3a, which range should be selected for the recorder? $\longrightarrow$

SAQ 5.3c
(cont.)

(*v*)   If we were to connect the pH-meter to the recorder detailed in SAQ 5.3a, is the performance of the recorder plus pH-meter system significantly worse than the performance of the pH-meter on its own?

If so, in which respect is it worse?

(*iv*) Is it possible with this instrument to correct the pH-meter if its sensitivity to pH has decreased due to ageing?

Summary

The final part brings together the topics introduced earlier and relates these to the final design and specifications of instruments.

The first section concentrates on spectrophotometer systems. These types of systems are used as very good examples which illustrate the application of many of the factors developed in the earlier parts of the unit. For a discussion of complete spectrophotometer instruments, the reader must refer to the appropriate units.

Gas chromatographs and pH meters are then used as examples of complete instruments to illustrate the idea that an analytical instrument contains an experimental process which is under the control of the operator of the instrument.

Finally the reader is introduced to the 'specifications' of the instrument as they might appear in a manufacturer's brochure. By using the knowledge gained in earlier sections, the performance of various instruments is predicted by examining their specifications.

Objectives

It is expected that, on completion of this Part, the student will be able:

- to predict the behaviour of spectrophotometer systems under given conditions,

- to predict the behaviour of a complete instrument under given conditions,

- to describe the use of computer memory systems to compensate for the effect of drift,

- to interpret the 'specifications' of various types of instruments in terms of their performance.

Self Assessment
Questions and Responses

SAQ 1.2a
A simple aneroid barometer is a measuring instrument designed to measure pressure (see Fig. 1.2e). It consists of flexible sealed bellows, which expand if the atmospheric pressure drops, and contract if the pressure rises. A mechanical lever-and-gear system operates a movable pointer which gives a reading of pressure.

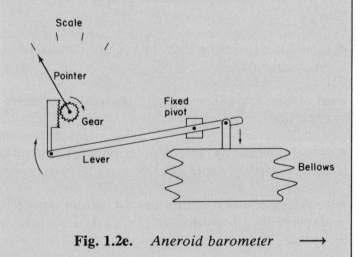

Fig. 1.2e. *Aneroid barometer* $\longrightarrow$

SAQ 1.2a (cont.)	Apply to this instrument the diagrammatic representation of a measuring instrument that was introduced in the text, and explain which parts of the barometer are represented by the following:

(*i*) transducer,

(*ii*) signal processor,

(*iii*) read-out system.

Response

The aneroid barometer *is* a basic measuring instrument. There are no electronic parts, but nevertheless we can still identify sections which relate to our basic diagram.

Starting from the 'input' end, the parameter being measured is the *pressure* of the atmosphere: that is the *input-signal*. The bellows converts the pressure (input-signal) into a mechanical movement (information-signal). The *bellows* is the *transducer*. The actual movement of the bellows is very small, and would be almost invisible to the eye. The purpose of the *levers and gears* is to amplify this movement: they are the *signal processor* section. At the 'output' end, the *needle-and-dial* is the visual display of the value of the pressure: the *read-out-system*.

It is not always easy to interpret any given measuring instrument in terms of a common diagram, and in fact it is only useful if it helps you to understand the instrument.

****

SAQ 1.3a   If the expression for the voltage from a pH elec-
                trode is rewritten as

$$E_V = E_o - A\,(\text{pH} - 7.00)$$

calculate:

(i) the value of A when the temperature of the
       solution is:

       (a) 20 °C, (b) 25 °C, (c) 30 °C;

(ii) the change in voltage if the pH of the solu-
       tion changes by one pH unit (eg from 3.00
       to 4.00) for each of the temperatures:

       (a) 20 °C, (b) 25 °C, (c) 30 °C.

Response

The full expression given in the text is

$$E_V = E_o - 0.198T(\text{pH} - 7.00)$$

and comparing this with the expression in the question,

$$A = 0.198 \times T.$$

(i) Thus for the three temperatures,

	Temp.	T	A
(a)	20 °C	293 K	58.0 mV
(b)	25 °C	298 K	59.0 mV
(c)	30 °C	303 K	60.0 mV

Note. You must remember that 'T' is in Kelvins.

(ii) To look at *changes* in E_V, we need to take differences

$$\Delta E_V = \Delta[E_o - A(\text{pH} - 7)]$$

At any fixed temperature, A will be constant.

$$\Delta E_V = \Delta E_o - A \times \Delta(\text{pH} - 7)$$

$\Delta E_o = 0$ since E_o is unchanged for changes in pH and temperature.

$\Delta(\text{pH} - 7) = \Delta(\text{pH})$ since '7' is a constant.

$$\Delta E_V = -A \times \Delta(\text{pH}).$$

or expressed in partial differentials,

$$\left(\frac{\partial E_V}{\partial(\text{pH})} \right)_T = -A$$

Hence, for a change of one pH unit, ie $\Delta(\text{pH}) = 1.00$

E_V changes by,

(a) -58 mV at 20 °C

(b) -59 mV at 25 °C

(c) -60 mV at 30 °C.

Thus, the signal processor in the pH-meter must change its amplification for different temperatures to allow for the change in the value of A. This is the reason that pH meters either have a manual 'set-temperature' control or automatic temperature compensation.

****

SAQ 1.3b

Assume that a filament in a TCD is made of a semi-conducting material, whose resistance decreases with increase of temperature. The filament is heated by the passage of a constant electric current, and pure helium gas flows over it. Some volatile organic material is added to the helium (without changing the flow rate). Indicate below whether the following parameters will increase or decrease.

Filament temperature: Increase / Decrease

Filament resistance: Increase / Decrease

Filament voltage: Increase / Decrease

Response

Helium gas has a *high* thermal conductivity. The mixture of helium gas, plus sample, will have a *lower* thermal conductivity that the pure gas. Thus, the heat will not be conducted away so quickly. The temperature of the filament will *increase*. In the question it states that the resistance of a semiconductor filament will *decrease* because of this *increase* of temperature. The variation in the voltage follows Ohm's Law:

$$V = I \times R$$

$$\Delta V = I \times \Delta R \text{ for a } constant \text{ current.}$$

Thus the *decrease* in resistance will give a *decrease* in voltage.

Note that for a *metal* filament (such as platinum), the resistance of the filament will increase if the temperature increases. The percentage change in resistance for the same change in temperature is smaller for a metal than for a semiconductor; but a metal filament can operate at a higher temperature and produces a more stable response over a greater range of experimental conditions.

****

SAQ 1.3c	List the Input-Signals and Output-Signals for each of the following:
	(*a*) photo-detector,
	(*b*) pH electrode,
	(*c*) TCD.

Response

	Transducer	Input-signal	Output-signal
(*i*)	Photo-detector	light intensity	electric current
(*ii*)	pH-electrode	concentration of H^+ ions	electric voltage
(*iii*)	TCD	concentration of sample in carrier gas	electric voltage

The answers to this question can be obtained easily by looking in the appropriate sections of the text. However, it is important to notice that the input signal to the light meter is in the form of *energy* actually following into the device, whereas the input signal to the other transducers is *information* about the value of some concentration in the system being measured.

You should also notice that it is not sufficient to say that the output is an electrical signal. We must be specific about the type of electrical signal – voltage or current. If a transducer gives an output in the form of an electric current, then it will require a signal processor different from that for a transducer which gives an electric voltage.

****

SAQ 1.4a

> By using the principal functional elements of an analytical instrument as headings in your answer, describe in note form the way in which a colorimeter works.

Response

Your answer to this question should have covered the points given below. The object of this question is to help you to think about what happens in each part of an instrument. Although most of the points listed are a re-iteration of the description given in Section 1.4.3, it is important that you should be able to re-group them under the appropriate headings. If you have problems with this then read Section 1.2 and Section 1.4 again, paying special attention to the concept of 'functional elements'. Additional comments are given below for further information.

Experimental Procedure.

Visible light from a simple bulb passes through a filter which selects a limited range of wavelengths (colour). The absorption of this light is measured by comparing the intensity of light passing through a 'blank' solution and a sample solution.

Control.

The operator can select which colour to use by choosing different filters. This depends on the colour of the solution.

The operator must calibrate each measurement by using a 'blank'. (In double-beam instruments this aspect of calibration is carried out automatically.)

The operator can adjust the amplification of the internal signal in the instrument so that with the blank solution, the meter in the colorimeter reads 100% *T*.

Analytical Variables.

Intensity of light is the quantity actually being measured within the instrument. (The percentage transmission of the light is the variable that the total experimental procedure is designed to measure.)

Detection System.

This is basically a simple light intensity measurement system.

Sample Preparation.

The sample solution must be available in a cuvette, together with a cuvette of the pure solvent. (With fairly concentrated solutions with an intense colour, the transmission of the light may be so small as to be very difficult to measure. It is then necessary to dilute the original solution, and to allow for this dilution in interpreting the results.)

Interpretation.

If the measurement is calibrated with the aid of a blank solution as described above, then the resultant transmission of the sample will be displayed directly on the meter. Any further interpretation of this information must be done by the analyst.

****

SAQ 1.4b   In what ways could an operator cause errors in measurement when using a melting-point apparatus?

An ingenious student is probably capable of finding many ways of misusing the instrument. However, for the purpose of this question consider the most probable errors, and associate these with particular functional elements given in Section 1.4.4.

Response

We cannot give a specific answer to this question, but we list some possible errors below as examples, together with the functional element with which they are associated.

One of the dangers is that the heating-rate is too fast. This may mean that there is a difference in temperature between the sample and the thermometer, thus giving a false reading. It may also be difficult to judge the exact temperature of melting if melting occurs too quickly – (control).

The thermometer may not be sufficiently accurate for the measurement – (detection system).

The operator may read the thermometer carelessly – (interpretation).

The sample may have become contaminated during preparation – (sample preparation).

You should be satisfied with your answer to this SAQ if you have been able to offer possible errors occuring within the functional elements of control, detection, sample preparation, and interpretation.

In almost all instrumental analytical measurements there are possible errors under each of these categories. You should look for them every time you use an instrument.

****

SAQ 1.5a	The term 'chromatography' covers a wide range of analytical procedures and instruments. Which of the following statements most clearly explains why it is possible to use the one name to cover all of the instruments? (*i*) The measurements can be made only on coloured solutions. (*ii*) They all employ broadly similar methods for measuring the amount of each component once they have been separated in the instrument. (*iii*) They all use broadly similar methods to separate the different components of the sample. (*iv*) The instrumental methods all relate to a limited range of chemical compounds.

Response

(*i*) It is true that 'chroma'tography was so called because the original experiments were done on coloured solutions. The original 'detector' was the chemist's eye!

However, most modern detectors do not rely upon the colour of the sample for detection (eg the TCD uses thermal conductivity). Hence statement (*i*) in the question is *false*.

(*ii*) Use of the term chromatography depends on the *experimental process* (see Fig. 1.4a) used to *separate* the components of the sample. The act of measuring the amount of sample is the function of the detection system (also see Fig. 1.4a), and is not part of the process which separates the constituents in the sample. Hence statement (*ii*) is also *false*. In fact there are many different types of detection systems for chromatography, which will be given in the appropriate units of the course.

(*iii*) In Section 1.5 we classified the main instrumental methods under four headings. The classification was based on a common experimental process being used in all methods within the group. In chromatography this method was the process of separating the various constituents of the sample before they reach the detector. Hence statement (*iii*) is *true*.

(*iv*) Classification of instrumental methods is based on the way in which the instrument operates, and not on the type of samples that can be analysed by the technique. This statement is *false*. It is true, however, that a particular experimental process may be applicable only to a limited range of chemical compounds, but the classification of the method is given by the experimental process and not by the samples being analysed.

****

SAQ 1.5b	What questions should you ask yourself when you are confronted with a new and unfamiliar analytical instrument that you wish to use?

Response

Before you can use an unfamiliar analytical instrument, there are many things that you need to know. However, it is essential that your answer should have included in some way the following three points.

What is the experiment being performed in the instrument?

What is being measured inside the instrument?

What is the operator able to do to change what happens inside the instrument?

The answers to these, or similar, questions should tell you basically what the instrument is doing! Once you understand what the instrument is doing you should also be asking 'how well does it do it?'. Your answer to the SAQ should therefore also include questions which ask what are the *capabilities* of the instrument, and what are its *limitations*.

****

SAQ 2.1a   Assume that the batteries in Fig. 2.1m have no internal resistance of their own, and calculate the current that flows through the 100 Ω resistor.

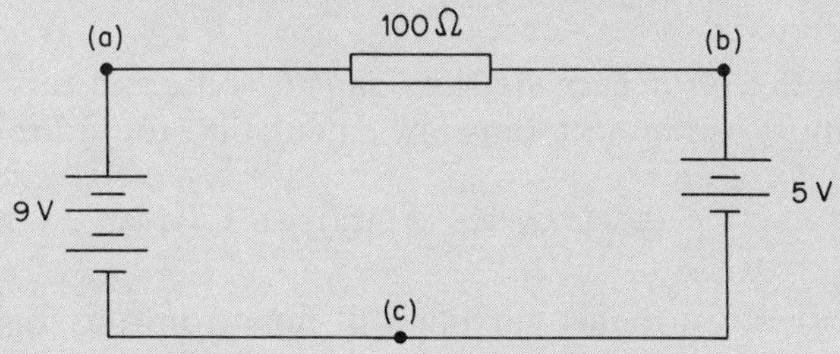

Fig. 2.1m

SAQ 2.1a
(cont.)

> Which way will the current flow through the re-sistor,
>
> (*i*)   from 'a' to 'b', or
>
> (*ii*) from 'b' to 'a' ?

Response

The 9-volt battery gives us the potential difference (pd) between 'a' and 'c':

$$V_{ac} = 9 \text{ V}$$

Similarly the 5 volt battery gives us the pd between 'b' and 'c':

$$V_{bc} = 5 \text{ V}$$

The pd between 'a' and 'b' is then given by,

$$V_{ab} = V_{ac} - V_{bc} = 9 - 5 = 4 \text{ V}$$

Thus the pd *across the resistor* is 4 volts, and this is the value that we must use in an Ohms law calculation to find the current.

$$I = V/R = 4/100 = 0.04 \text{ A} = 40 \text{ mA}$$

The (conventional) current will flow from the higher to the lower potential. Thus this is (*i*) – from 'a' to 'b'.

If you have used a different value for V in the calculation then it may be because you do not understand that it is the potential difference actually *between the ends* of the resistance that must be used, not just the potential at either end (see Section 2.1.3).

****

SAQ 2.1b

A rotary 'pot' as in Fig. 2.1n has a circular resistance track of uniform resistance forming an arc of total angle 270 degrees. A wiper makes electrical contact to the track and can be turned so that the angle θ varies between 0 and 270 degrees.

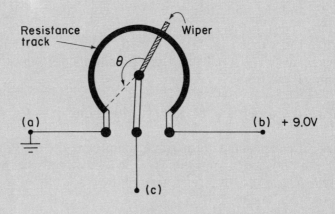

Fig. 2.1n

The terminal 'a' is connected directly to 'earth' and terminal 'b' connected to a potential of $+9.0$ V.

Calculate the angle θ required for the wiper if the centre terminal, 'c', is to have a potential of 3.0 V.

Response

For the theory on this question refer to Section 2.1.6, and the 'potential divider'.

The two sections of the resistance track on either side of the wiper can be considered as the two resistors, R_1 and R_2, in the potential divider in Fig. 2.1.f(*ii*).

The fact that 'a' is connected to 'earth' is another way of saying that the potential at 'a' is zero, ie

$$V_a = 0$$

We are also told that the potential at 'b' is 9 V, $V_b = 9$ V.

Thus referring to Fig. 2.1.f(ii) the input-potential V_b is equivalent to V_i, and the potential at 'c', V_c, is equivalent to V_o.

When the wiper is at 'a', ie $\theta = 0°$, then the wiper is directly connected to 'a' ($R_1 = 0$) and the potentials at 'a' and 'c' will be the same, $V_c = 0$ V.

When the wiper is a 'b', ie $\theta = 270°$ then $V_c = V_b, = 9$ V.

Since the resistance of the track is uniform, the resistance (R_1) of the section between 'a' and the wiper will be directly proportional to the angle, θ, and the total resistance of the track ($R_1 + R_2$) proportional to the total angle, 270°.

Then,

$$V_c = V_b \times R_1/(R_1 + R_2) = V_b \times \theta/270$$

Thus,

$$3.0 = 9.0 \times \theta/270$$

$$\theta = 90°.$$

If your answer is 180° then you have taken the wrong point for the zero potential in this problem. Since 'a' is at zero potential, the potential at 'c' is given by the difference in potential between 'a' and 'c' and *not between* 'b' and 'c'.

****

SAQ 2.1c Calculate the AC current required for a 3 kW water-bath heater to be run from the 240 V mains supply.

State whether your value of the current is the 'peak' value or the 'root-mean-square' value.

Response

The value of 240 V for the mains voltage can be used directly for calculations for power – see the question in Section 2.1.9.

Since

$$\text{Power, } W = V \times I,$$

we have

$$\text{Current } I = 3000/240 = 12.5 \text{ A}$$

In AC electricity it is the 'rms' value that is used in calculating power (see the definition of 'rms' in Section 2.1.9), hence 240 V is the root-mean-square value of the mains voltage

****

SAQ 2.2a Does the battery for a digital watch supply

(*i*) power,

(*ii*) information,

(*iii*) a combination of both power and information?

Please mark one of these answers.

Response

The battery in a digital watch supplies only power to drive an electronic 'chip'. It does not provide any information about the passing of time. The timing process comes from within the 'chip'.

****

SAQ 2.2b

Indicate which parameter(s) below may convey information in the DC, the AC, and the digital mode.

1. Voltage.   2. Current.   3. Frequency.

4. Polarity.   5. Coded pulses.   6. Phase.

Write the appropriate numbers (1–6) alongside each of the following:

DC

AC

Digital

Response

The correct answer is:

DC   – voltage, current, polarity   (1, 2, 4)

AC   – voltage, current, frequency, phase   (1, 2, 3, 6)

Digital – coded pulses   (5)

It is possible to construct the above answer by using several statements, viz:

(*i*)   The magnitude (voltage, current) of the signal is a variable parameter for both DC and AC (see Section 2.2.6 and Section 2.2.7);

(*ii*) Polarity can apply only to a DC signal (see Section 2.2.6);

(*iii*) Frequency and phase apply only to an AC signal (see Section 2.2.7).

(*iv*) The magnitude (voltage, current) of digital pulses is not a variable parameter – only the 'coding' of the pulses (see Section 2.2.8).

If you have not got the correct answers then check which of the above statements you are unsure about, and then refer back to the appropriate parts of the text.

$$***$$

SAQ 2.2c | The radiation source in a certain simple spectrophotometer is a 12 V (DC), 24 W, tungsten lamp. This is supplied by a circuit similar to that in Fig. 2.2a. Calculate the minimum current that must be drawn from the 240 V mains supply.

Hint: You need be concerned only with the conservation of energy.

Response

Conservation of energy requires that the average flow of energy into the instrument (input power) from the mains must not be less than the flow of energy (output) required to drive the lamp.

The power of the lamp is 24 W, thus at least 24 W must be drawn from the mains. Since $P = V \times I$ we can calculate the minimum current.

$$I = P/V = 24\text{W}/240\text{V} = 0.10 \text{ A}.$$

This is the minimum (rms) current required to drive the instrument. Of course, there will be other components requiring power in the instrument, resulting in a higher actual requirement than 0.10 A. However, production of heat, light, or motion (in that order) is the most 'power-hungry' part of an instrument.

Handling of information, on the other hand, can nowadays be done with a very minimum of power, for eg: digital watches with a *five-year* battery life!

****

SAQ 2.2d   Refer to Fig. 2.2g, and calculate:

(*i*)   the frequency of each wave,

(*ii*) the phase difference between waves S and R,

(*iii*) the DC voltage that would be obtained from a PSD with $K = 1$, (see Section 2.2.7) if S is the signal wave and R is the reference wave.

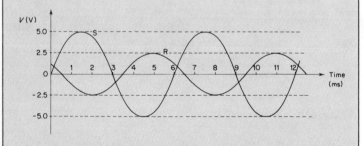

Fig. 2.2g. *Wave forms for use in SAQ 2.2d*

Response

The period, T of each wave, S and R, is the same. One cycle of each wave takes 6 ms. If you have calculated that one cycle is equal to 3ms. then you have taken only one half of the cycle.

A complete cycle must include all 'phases' of the wave once, returning to the same phase as the starting-point. The most accurate way to measure the length of a cycle is to measure the difference in time between respective zero-crossing points of the same phase. A 'zero-crossing point' is where the wave crosses the zero line positive to negative or *vice versa*. If the wave goes from positive to negative (negative-going) then this is at a different 'phase' of the wave from a zero-crossing point when the wave goes from negative to positive (positive-going).

Thus $T = 6$ ms, and the frequency, $f = 1/T = 167$ Hz.

In order to measure phase difference, we must measure the relative displacement of the two waves along the time axis, and to do this we measure the difference in time between respective zero-crossing points of the same phase.

The R wave has a positive-going zero-crossing point at $t = 3.5$ ms, and the nearest equivalent crossing point for S is at $t = 6.0$ ms.

Thus we can say that R 'leads' S by a phase angle, θ given by,

$$\theta = 360 \times \text{(difference in time)/(period of wave)}$$

$$= 360 \times (6 - 3.5)/6 = 150°.$$

Then, to use the expression given in Section 2.2.7 for the PSD, we must first measure the peak value, V_{ac} of the signal wave, S.

$$V_{ac} = 5 \text{ V}.$$

Note that the magnitude of the reference wave R does not enter into the calculations.

Then

$$V_{dc} = K \times V_{ac} \times \cos \theta = 1 \times 5 \times \cos 150° = -4.33 \text{ V}.$$

If you have got a different value for the phase difference, then it may be that you have taken the difference in time between points on S and R which are not in the same phase – recheck this.

Alternatively, you may have obtained a value of 210°, which is essentially the same as above since $210 = 360 - 150$, and it gives the same result (as it should) when the waves pass through the PSD.

****

SAQ 2.3a

Many instruments have an *analogue output* giving a DC electric signal which can be fed into some other measuring instrument such as a chart recorder. Assume that such an output is giving a DC voltage of 60 mV on open circuit (ie with nothing connected), and that the output impedance of this source is 1 kΩ.

(*i*) Calculate the actual voltage that would be measured if a chart recorder of input impedance 100 kΩ is connected to the output.

(*ii*) Express the difference between the two voltages as a percentage of the open-circuit voltage.

Response

In order to answer this question we shall assume that the source behaves as a perfect Thévenin equivalent circuit (see Section 2.3.2).

The next step is to write down the values of E and R_o. The value of R_o has been given explicitly as 1 kΩ.

We can also derive the value of E because we are told that the 'open-circuit' voltage of the source is 60 mV. We have already seen in Section 2.3.3 that the open-circuit voltage, V_t, is equal to the emf, E. Thus $E = V_t = 60$ mV.

We must represent the *effect* of connecting the chart recorder to the analogue output of our instrument by connecting a resistance (R_i) of 100 kΩ to our Thévenin circuit. To calculate the actual voltage measured, V, we perform a similar calculation to that used in Section 2.3.2. The current flowing from the source will be given by:

$$I = 60 \text{ mV}/(100 + 1) \text{ k}\Omega$$

$$= 0.594 \ \mu A$$

The voltage actually measured is that voltage which appears across the external resistance, hence;

$$V = 0.594 \ \mu A \times 100 \text{ k}\Omega$$

$$= 59.4 \text{ mV}.$$

The actual error is $(60 - 59.4)$ mV $= 0.6$ mV, and the percentage error is:

$$\text{Error} = 0.6/60 \times 100\% = 1\%$$

From this calculation it is possible to see that an approximate 'rule of thumb' for *small* errors, which arise because the measuring instrument does not have an infinite input impedance, is as follows.

$$\text{Error} = R_o/R_i \times 100\%$$

****

SAQ 2.3b   Assume that the thermocouple used as an ir detector (see Section 2.3.6) has an output impedance of 10 Ω and that, with a given intensity of ir radiation, it is producing an emf of 7.0 μV.

Calculate, for the same intensity of IR radiation,

(i)   the maximum voltage signal,

(ii)   the maximum current,

(iii)   the maximum power,

that would be available if it were possible to change the input impedance of the signal processor, to which the output thermocouple is connected.

Response

Again we use a Thévenin equivalent circuit. Notice how we are able to use this same circuit for different types of transducers and also to represent other electrical outlets as in the previous SAQ, 2.3a.

In this case $E = 7.0$ μV, and $R_0 = 10$ Ω

The three parts of the question correspond to the three main 'conditions' given in Section 2.3.6 for connecting an electrical source to a receiving circuit (in this case the signal processor).

To obtain these three conditions we need to change the input impedance, R_i the signal processor circuit.

(i)   For maximum voltage we put $R_i = \infty$, then $V = E = 7.0$ μV (*ii*) For maximum current we put $R_i = 0$, then $I = E/R_0 = 0.7$ μA

(*iii*) For maximum power we put $R_i = R_0$, then $P = E^2/4R_0 = 1.225 \times 10^{-12}$ W.

****

SAQ 2.4a

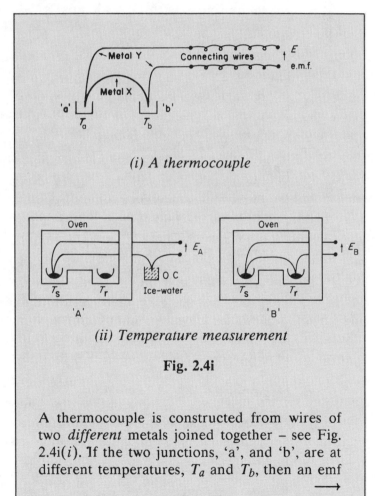

(i) A thermocouple

(ii) Temperature measurement

Fig. 2.4i

A thermocouple is constructed from wires of two *different* metals joined together – see Fig. 2.4i(*i*). If the two junctions, 'a', and 'b', are at different temperatures, T_a and T_b, then an emf

$\longrightarrow$

SAQ 2.4a (cont.)

(voltage) will be produced in the wire. This emf, E, depends on the *difference* between T_a and T_b, and may often have a *linear* relationship of the form:

$$E = k(T_a - T_b)$$

where k is a constant of the thermocouple. For the purposes of this question use 40 μV $°C^{-1}$ as a typical value for k, ie a temperature difference of 70 °C would produce an emf of 70 × 40 μV = 2.8 mV.

In Differential Thermal Analysis, DTA, a sample is slowly heated in some form of oven or heating block. At the same time, some reference material is also heated in a similar way. Normally the rate at which the temperature of the sample rises will be uniform, except where the sample undergoes some physical or chemical change. This change may be accompanied by the absorption of a large amount of heat or even by the release of heat due to an exothermic reaction. The rate of temperature rise of the reference material would not be affected by such variations. An example is given in the table below of a typical variation in the temperatures of a sample, T_s, and reference material, T_r, when heated in an oven at a controlled rate.

Consider the two situations depicted in Fig. 2.4i(ii). In each situation one junction of a thermocouple is placed in the sample, but the other junction is placed as in the two situations below.

'Situation A', the other junction is placed in an ice-water mixture to maintain a constant temperature of 0 °C: thermal analysis. $\longrightarrow$

SAQ 2.4a
(cont.)

'Situation B', the other junction is placed in the reference material at the varying temperature, T_r: differential thermal analysis.

(*i*) Plot on the two graphs provided [Fig. 2.4j(*i*) and (*ii*)] the variation of the emf's produced in the two situations over the complete temperature range. Use $k = 40$ μV °C^{-1} as above.

Note the *different abscissae* for the two graphs.

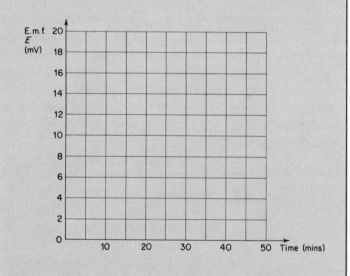

(i) Situation 'A'

Fig. 2.4j $\longrightarrow$

SAQ 2.4a
(cont.)

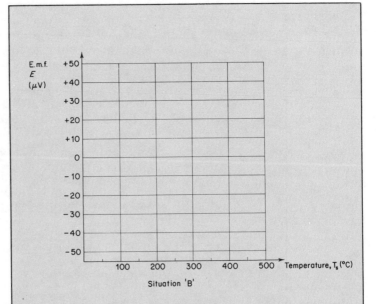

(ii) Situation 'B'

Fig. 2.4j (cont.)

(*ii*) Calculate, from the graphs, the average rate, in °C min^{-1} at which the temperature of the oven is rising.

(*iii*) Identify points where a fluctuation occurs in the heating curve of the oven.

(*iv*) Determine at what temperature some form of physical or chemical change occurs in the sample.

(*v*) Comment on the advantages of placing one junction of the thermocouple in the reference material. $\longrightarrow$

SAQ 2.4a
(cont.)

Table of results.

Time (min)	T_r (°C)	T_s (°C)
0	50.0	50.0
5	100.0	100.0
10	150.0	150.0
11	160.0	160.0
12	170.0	169.5
13	180.0	179.0
14	195.0	194.5
15	210.0	210.0
17	240.0	240.0
19	265.0	265.0
21	280.0	280.0
23	290.0	290.0
25	300.0	300.0
27	320.0	320.0
29	340.0	340.0
31	355.0	355.25
32	360.0	360.5
33	365.0	365.25
34	370.0	370.0
36	385.0	385.0
38	410.0	410.0
40	440.0	440.0
42	470.0	470.0
44	490.0	490.0
45	500.0	500.0

Response

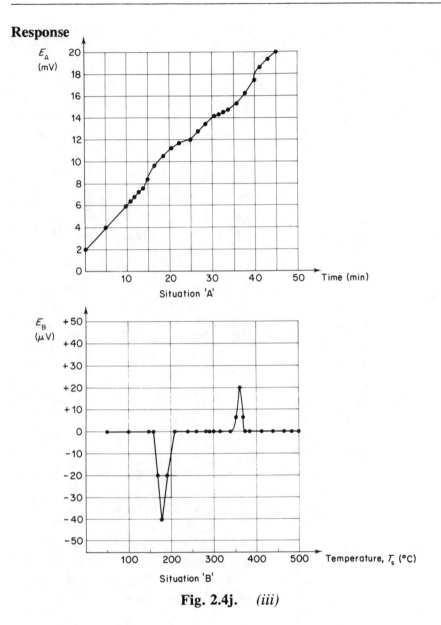

Fig. 2.4j. *(iii)*

(*i*)   The two graphs are plotted in Fig. 2.4j(*iii*).

In 'A', the emf from the thermocouple increases from 2 mV (equal to $50 \times 40 \ \mu V$) to 20 mV, and both the graph, and the input to the amplifier, must be able to accomodate this range of voltages. In 'B' however, the use of a comparative method allows us to concentrate *only* on the emf that occurs as a result of the slight differences between the sample and the reference material, and the gain of the amplifier could be increased to respond the very small emfs as in 'B' without being swamped by the large emfs that occur in 'A'.

(*ii*) By drawing a 'best straight line' through the points on graph 'A' we can see that the emf increases by an amount,

$$\Delta E = 20 - 2 = 18 \text{ mV}$$

This can be converted into a temperature interval, ΔT, by using the value of k for the thermocouple

$$\Delta T = \Delta E / k = 18000 \, (\mu \text{ V}) / 40 (\mu \text{ V}.\,^{\circ}\text{C}^{-1}) = 450 \,^{\circ}\text{C}.$$

This occurs over a time interval, $\Delta t = 45$ min.

Thus the mean rate of temperature increase is

$$\Delta T / \Delta t = 450/45 = 10 \,^{\circ}\text{C min}^{-1}$$

If your answer is 11.1 $^{\circ}$C min^{-1}, then you have not allowed for the fact that the initial temperature is 50 $^{\circ}$C and not 0 $^{\circ}$C.

(*iii*) Fluctuations in the heating curve of the oven can be seen from graph 'A'.

An upward fluctuation occurs between 190 $^{\circ}$C and 290 $^{\circ}$C reaching a maximum deviation of $+25$ $^{\circ}$C.

A downward fluctuation occurs between 350 $^{\circ}$C and 450 $^{\circ}$C reaching a maximum deviation of -25 $^{\circ}$C.

If you have not obtained these answers then it is possible that you have made an error in drawing the graph.

(*iv*) The occurrence of some physical or chemical change in the sample will be observed as a slight fluctuation in its heating curve. It is quite clear that by using graph 'A' it would be very difficult to detect the very slight fluctuations that occur in the heating curve for 'T_s'. However, if we use graph 'B' then these slight effects can be noticed quite clearly.

One event occurs with a peak at 179 °C, with a magnitude of -1 ° C. The temperature of the sample is less than the reference, which implies that some of the heat from the oven has been absorbed by the sample in some change or reaction instead of being used to raise the temperature of the sample.

Another event occurs at 360 °C, with a magnitude of $+\frac{1}{2}$ °C. Here T_s is higher than expected, which must mean that the sample is generating heat, ie an exothermic reaction takes place at 360 °C.

An important analytical parameter in this measurement is the temperature at which the event occurs. Hence the abscissa for graph 'B' is the temperature of the sample, T_s.

(*v*) In the example given here there are two main effects – a fluctuation in the heating curve of the oven and some change associated with the sample. We can see from (*iii*) and (*iv*) that the fluctuations in the temperature of the oven ($+25$ °C) are considerably greater than the analytical signal (-1, $+\frac{1}{2}$ °C), and that, only by using the reference material for one junction of the thermocouple, can we eliminate effects other than the analytical signal. Using the reference material in this way, allows us to perform a 'comparative' measurement.

It enables us to measure the *signal* that we want in the presence of a lot of *noise* due to fluctuations in heating rates. In fact, if you look back at the results and calculations, you will see that the oven fluctuations partially *overlap* the analytical signals, and without this differential method the analytical signals would be extremely difficult to distinguish from other variations.

It would be a useful exercise for you to compare the DTA instrument outlined here with the Melting-Point Apparatus described in Section 1.4.2.

****

SAQ 2.4b   Assume that a circuit as in Fig. 2.4g, with semi-conductor filaments, is used for a TCD system, and that, for a certain flow-rate of helium carrier gas, the two voltages, V_a and V_b, are equal in the absence of any sample in the gas. Indicate below the way in which these two voltages may change if the flow-rate of helium *decreases*. (Note that this question is concerned only with a *change* of *flow-rate* and not with the presence of any sample in either of the two gas flows.)

If you are in any doubt about the behaviour of the TCD detector, refer back to Section 1.3.3. Note that the positive potential is being applied to the *top* of the bridge in the diagram.

Possible responses.

(i) V_a increases and V_b decreases,

(ii) V_a increases and V_b increases,

(iii) V_a decreases and V_b decreases,

(iv) V_a decreases and V_b increases.

Response

If the flow rate of the helium decreases, then the heat energy of the filament will not be conducted away so quickly. Thus the temperature of the filament will rise. A semiconductor filament has a negative temperature coefficient of resistance, ie if the temperature rises the resistance will fall. Since both filaments are subjected to the same gas flow, the resistances of each of them will decrease. It might be helpful, if you have difficulties with this part of the question, to refer back to SAQ 1.3b.

Having established that the filament resistances decrease, it is then a matter of finding what happens to each of the potentials, V_a and V_b, in the resistor chains when the upper resistance is reduced. Referring to Section 2.4.6 and Fig. 2.4f, if R_2 is reduced then V_a increases. Similarly if R_4 is reduced the V_b increases. The correct answer is therefore response (*ii*). Note that the output from the bridge is the *difference* between V_a and V_b $(= V_a - V_b)$, and that, if *both* V_a and V_b increase by the same amount, there will be no effect on the output from the bridge due to a change in flow rate.

If you have chosen response (*iii*), then it may be that either you thought that the resistances should increase, in which case look back to Section 1.3.3, or you have made an error in the use of the equation for the potential divider – see Section 2.1.6 and Section 2.4.6. In either event, this is not a particularly serious error as you have realised that both V_a and V_b change in the *same* way.

If you have chosen either of the responses (*i*) or (*iv*), then you have missed the crucial point of having *two* detectors, one for the sample and one as a control or reference. In this way any variation in flow-rate affects both detectors similarly and will be cancelled out by making a comparative measurement. Check back with Section 1.3.3.

****

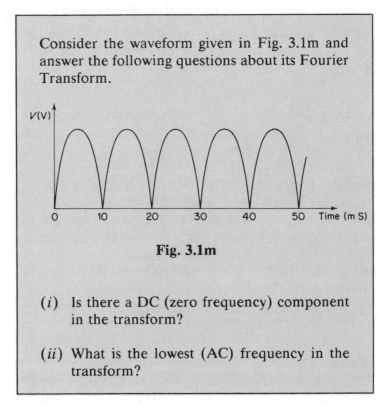

SAQ 3.1a   Consider the waveform given in Fig. 3.1m and answer the following questions about its Fourier Transform.

Fig. 3.1m

(*i*)   Is there a DC (zero frequency) component in the transform?

(*ii*) What is the lowest (AC) frequency in the transform?

Response

You may have recognised the waveform in the question. It first appeared in Fig. 2.2b(*ii*), when we were dealing with the conversion of AC power into DC. (Not surprisingly we shall find that there *is* a DC component.)

(*i*) In order to identify a DC component we must consider what would be the value of the voltage if we were to take an *average* over one complete cycle. The magnitude of this average gives the amplitude of the DC component. In this example, the average cannot be zero because the voltage never becomes negative. There must be a DC component.

To calculate the fundamental (AC) frequency, we must look at the time taken for each cycle – the period. T. This can be seen to be 10 ms. The equivalent frequency is then $f = 1/T = 100$ Hz – (see Fig. 3.1o).

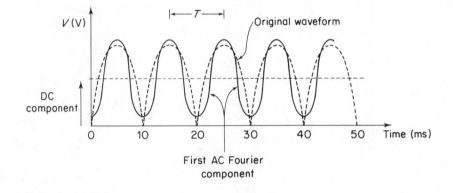

Fig. 3.1o

Of course, higher frequency components will be required to give the exact *shape* of the wave, but we have found the two most significant components of the waveform.

I suggest that you refer back to Fig. 2.2b. In that you will see that the effect of adding the capacitor is greatly to reduce the fundamental AC component to give a smoother DC voltage. In instruments this AC component on the DC power supply is called *mains ripple*, and may occur (in the UK) at 100 Hz or 50 Hz depending on the type of rectification.

Battery-driven instruments do not suffer from mains ripple.

****

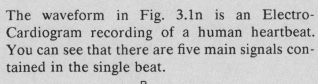

SAQ 3.1b   The waveform in Fig. 3.1n is an Electro-
Cardiogram recording of a human heartbeat.
You can see that there are five main signals con-
tained in the single beat.

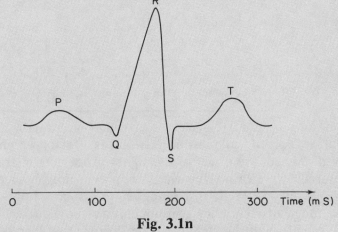

Fig. 3.1n

Estimate the bandwidths of the frequencies contained in each of the five signals.

What do you think would happen, in general
terms, to the shape of the heart-beat signal if
it were processed by an instrument which could
cope only with a bandwidth of 30 Hz?

Response

The scale in Fig. 3.1n is already expressed directly in time (ms), thus
we can go directly to measuring the 'widths at half-height' for each
of the signal elements. We can then use the Eqn. (3.5) from the text
to estimate bandwidths, Δf. If we refer to Fig. 3.1n, this gives the
following results

Signal Elements	Approximate Width at Half Height/ms	Approximate Bandwidth/Hz
P	30	17
R	30	17
T	30	17
Q	6	80
S	4	125

The last part of the question introduces a factor which we have not yet discussed, the effect of the signal processing instrument on the observed waveform. However, I hope you might have guessed the correct answer. If the instrument cannot successfully cope with the full range of frequency components in the Fourier Transform of the signal, then the signal will be distorted or it may even be lost altogether. Here the bandwidth of the instrument can successfully pass the signal elements P, T and R, but it would severely attenuate the high-frequency elements, Q and S.

****

SAQ 3.2a

A particular type of small transistor radio has an output-amplifier system with a frequency response from 150 Hz to 6.00 kHz.

(*i*)   Is this an AC or a DC amplifier?

(*ii*) Calculate the band-width of the amplifier.

(*iii*) Explain why the small band-width gives poor quality sound.

Response

(*i*)   6 kHz is the high-frequency cut-off, f_h = 6 kHz. 150 Hz is the low-frequency cut-off, f_l = 150 Hz. A DC amplifier has a frequency response down to zero frequency (f_l = 0). Hence this is not a DC amplifier. It is an AC amplifier.

(*ii*) The band-width Δf = $f_h - f_l$ = 5.85 kHz.

(*iii*) The question implied that the band-width of the radio was small. In fact, the band-width of the adult human ear is from about 20 Hz to 15 kHz (see Section 3.2.3) and the ear can hear far more frequency components in any given signal than the radio is able to convey. Thus the radio removes components from the Fourier Transforms of a sound, thereby reducing the quality.

****

SAQ 3.2b

A block diagram of a simple pH meter is given in Fig. 3.2e. Two amplifiers A and B are shown.

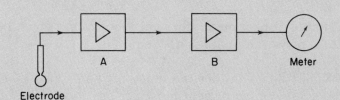

Fig. 3.2e

The input to A from the pH electrode is a voltage signal which changes by 59.0 mV for unit change on the pH scale. The output from A also changes by 59.0 mV per pH unit. $\longrightarrow$

SAQ 3.2b
(cont.)

The output from B operates the display meter and is a voltage signal which changes by 1.00 volt for unit change on the pH scale.

(i) State which type of amplifier should be used (a) for A, and (b) for B.

Choose from: AC amplifier, DC amplifier, Buffer amplifier, Power amplifier, Logarithmic amplifier.

(ii) Based on the information in the text, estimate the voltage gain for each of the amplifiers A and B.

Response

(i) Let us first eliminate some of the possibilities. The signal in a pH meter is definitely DC, hence there are no AC amplifiers.

The derivation of pH value from the pH-electrode voltage involves a linear relationship. There is no need of a log amplifier.

A simple display-meter does not require very much power to operate. A power amplifier is not required.

We must now consider the functions performed by the pH meter. The complete instrument must perform *two* functions: it must respond to the voltage from the pH-electrode while only drawing a minute current, and it must amplify the signal from 59.0 mV to 1.00 volt. We must therefore have a *buffer* amplifier and a *DC voltage* amplifier.

A must be the buffer amplifier so that only a minute current is drawn from the pH electrode. This is consistent with the fact that we are told that the gain of A is unity: the input and the output voltage are both 59.0 mV pH^{-1} unit. Hence B must be the DC voltage amplifier.

If you made A the DC voltage amplifier, the effect would be to draw too much current from the pH-electrode. The *buffer* amplifier must come first.

(*ii*) The voltage gain of the buffer amplifier is 1.

The voltage amplifier, B, must increase 59.0 mV to 1.00 V (= 1000 mV), ie a gain G of $= \dfrac{1000}{59.0} = 16.95$.

****

SAQ 3.3a

A signal as in Fig. 3.3j consists of a DC voltage and an AC 'ripple'. This type of voltage may exist in the DC power supply for an instrument, see Section 2.2.4. If we wish to measure the amplitude of the AC ripple, V_{ac}, we may find problems because of the large DC voltage associated with it.

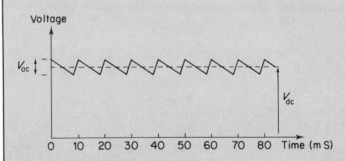

Fig. 3.3j

We can however pass this signal through a filter so that we have only the AC parts of it. Describe the characteristics of a filter which will do this. Estimate the value of the high or low cut-off frequency or band-width as appropriate for the type of filter you have chosen.

Response

The component we wish to remove is the DC (zero frequency) component, but we wish to retain all relevant AC components.

The AC component with the lowest frequency is that of 100 Hz, thus we must choose a high-pass filter with a low frequency cut-off below 100 Hz.

Thus the correct answer to the problem is a *high-pass filter* with a low-frequency cut-off, f_l where 0 Hz $< f_l <$ 100 Hz.

If you have chosen a low-pass filter or a notch filter, then you have misunderstood the problem. It is the very low frequency, V_{dc}, that we do not want to pass. If, however, you chose a band-pass filter then you would certainly eliminate the DC component, but you would also eliminate any AC frequency component greater than f_h. This may have the effect of distorting the AC signal itself. A tuned filter would not be appropriate for the same reasons as for a band-pass filter. The distortion would, in fact, be worse as only a few AC components could pass through the tuned filter.

****

SAQ 3.4a   In the diagram of an optical detection system in Fig. 3.4g, the light beam is 'chopped' by a disc rotating at 50 rotations per second. The disc has 4 segments as shown: two segments block the light completely and two transmit the light.

The intensity of light is converted into an electrical signal by a photo-detector. This signal is to be amplified by a tuned amplifier. $\longrightarrow$

**SAQ 3.4a
(cont.)**

At what frequency should the amplifier be tuned?

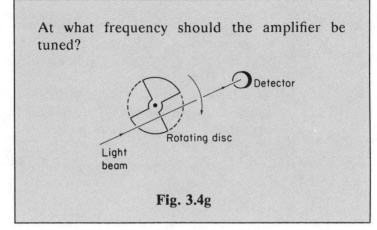

Fig. 3.4g

Response

During each rotation of the disc the light signal will go from a maximum to a minimum *twice*. This is equivalent to two complete cycles. Thus if the frequency of rotation of the disc is 50 Hz then the effective modulation frequency of the light is $2 \times 50 = 100$ Hz.

The electrical signal will be a square wave with this frequency. We would normally have the amplifier tuned to the same frequency, ie $f_0 = 100$ Hz.

However it may be better, for various reasons, to tune the amplifier to some higher-frequency harmonic component in the signal – eg 3 $\times$ f_0 for a square wave.

****

SAQ 3.4b
A block diagram, Fig. 3.4h, represents a system which amplifies a DC signal by using a modulation technique, followed by phase-sensitive detection. However, too many information pathways have been drawn in the diagram. Only *one* of the paths, A, B, or C, should be kept in the diagram. Which *one* pathway is essential? Choose from A, B, and C.

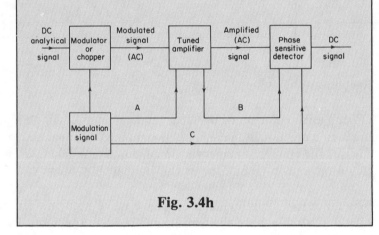

Fig. 3.4h

Response

We know that a PSD must have a *reference* input as well as the AC *signal* input. this tells us that either B or C must be retained as an essential connection.

We also know that the reference input to the PSD must contain timing information concerning the phase of the modulation process. This must come along path C.

∴   It is only path C that must be retained.

If you also look back to the sections on filters and amplifiers you will see that we described tuned amplifiers as requiring only an input and an output for the signal. No other information was required to flow into or out of the circuit. This confirms the fact that the pathways A and B are not appropriate in this diagram.

****

SAQ 3.5a	Explain the fundamental difference between most noise signals and analytical signals, apart from the simple fact that we do not want the noise but we do want the analytical signal.

Response

The answer to this question is very short. Most noise signals are random but an analytical signal must have some regular pattern.

You may have tried incorrectly to explain the difference in terms of the shape of the Fourier Transform, eg $1/f$-noise or white noise. In this you would be wrong because it is conceivable that an analytical signal could have a similar shape for its Fourier Transform. The difference is that the *components* of a noise signal are not related to one another whereas the components of an analytical signal maintain a constant relationship, giving a regularity in the signal.

The exception to randomness in noise occurs from environmental noise. A good example is 50 Hz pick-up which will appear in a circuit with a well-defined pattern and, as such, will be indistinquishable (except perhaps in phase) from a 50 Hz analytical signal.

****

SAQ 3.5b   Decide which of the following statements are
true and which are false.

(*i*)   Johnson noise in a circuit can be eliminated by cooling the circuit to 0 °C.

(*ii*) White noise occurs only at optical frequencies.

(*iii*) White noise has equal components at all
frequencies.

(*iv*) White noise can be eliminated by using appropriate filters.

(*v*)   Modulation of a DC signal can be used to
reduce $1/f$-noise.

Response

The answers are:

(*i*)   F,

(*ii*)   F,

(*iii*) T,

(*iv*)   F,

(*v*)   T.

(*i*)   If your answer was 'true' then you have probably confused the
Centigrade and Absolute scales. In the equation for Johnson
noise the temperature is Kelvins.

(*ii*) The term 'white' is used only as an *analogy* to the fact that
white light has frequency (wavelength) components across the

complete visible spectrum, whereas 'white noise' covers all electronic frequencies starting from 0 Hz and theoretically extending to infinite frequency.

(*iii*) If you failed to get the correct answer here you should refer back to the answer in Section 3.5.3.

(*iv*) White noise can *never* be totally eliminated. Its effect can be reduced by appropriate filters, but some components of it will always be present.

(*v*) If you failed to get the correct answer here you should refer back to the answer in Section 3.5.4.

****

SAQ 4.1a

Complete the following sentences by using phrases from the list given below.

(*i*) .... can be a serious problem in an integrating system.

(*ii*) The output from a differentiator circuit can easily suffer from ....

Possible phrases:

(*a*) wavelength scan,
(*b*) attenuation,
(*c*) high-frequency noise,
(*d*) AC drift,
(*e*) amplification,
(*f*) DC drift.

Response

The correct statements are as follows.

(*i*) *DC drift* can be a serious problem in an integrator system.

(*ii*) The output from a differentiator circuit can easily suffer from *high-frequency noise*.

If you have used the correct phrases but have put them in the wrong sentences, then you should check in the text to see if you have made a simple slip. If you are still confused it means that you basically misunderstand the application of the Fourier spectrum and the gain characteristics in Fig. 4.1a to integration and differentiation. If so, you must work steadily through the text again and perhaps seek other advice.

If you have used 'AC drift' then you need to look back to Section 3.2.4 to understand the use of the term 'drift'.

'Wavelength scan', 'amplification' and 'attenuation' are all factors that you may indeed have to consider in integration and differentiation circuits. However, the real *problems* in using an instrument are involved in reducing the various types of drift and noise associated with the signal. Hence we obtain the answers given above.

****

| SAQ 4.1b | With reference to a wavelength spectrum which contains a broad line, A, and a narrow line, B, $V(A)/V(B)$ is the ratio of the magnitudes of the two lines in the normal spectrum, and $V(2)(A)/V(2)(B)$ is the ratio of the magnitures of their second derivatives with respect to time. $\longrightarrow$ |

SAQ 4.1b
(cont.)

Which of the following statements are true?

(i) $V(2)(A)/V(2)(B) < V(A)/V(B)$ T / F
(ii) $V(2)(A)/V(2)(B) > V(A)/V(B)$ T / F
(iii) $V(2)(A)/V(2)(B) = V(A)/V(B)$ T / F
(iv) $V(2)(A)/V(2)(B)$ will increase if the wavelengths are scanned more quickly.
(v) $V(2)(A)/V(2)(B)$ will decrease if the wavelengths are scanned more quickly.

Response

The second derivative emphasises lines with a narrow linewidth, here B. Thus $V(2)(B)$ will be 'amplified' more than $V(2)(A)$, and we would have:

$$V(2)(B)/V(2)(A) > (V)(B)/(V)(A)$$

This is equivalent to relation (i). Thus relations (ii) and (iii) are incorrect.

If the lines are scanned more quickly then both $V(2)(A)$ and $V(2)(B)$ will increase. However the proportional increase is the same for both, and the ratio $V(2)(A)/V(2)(B)$ will not be changed by changing the sweep rate. Therefore statements (iv) and (v) are *both* incorrect.

The only correct statement is (ii).

If you are in doubt about this, refer back to the last exercise and the lines A and B in that example, and use the values of $V(2)(A)$ and $V(2)(B)$ obtained in the answer to check the validity of the statements in this SAQ.

****

SAQ 4.2a   Assume that, when we are recording the energy
spectrum of a beam of γ-rays with a multichan-
nel analyser, we decide to use 1024 channels instead of 512 for the same range of γ-ray energies
and for the same counting period.

Which of the following statements are true?

Increasing the number of channels means that:

(*i*)   it is easier to distinguish between photons
of nearly the same energy,

T / F

(*ii*)   the signal-to-noise ratio of the spectrum is
increased,

T / F

(*iii*) the signal-to-noise ratio of the spectrum is
reduced,

T / F

(*iv*) we can halve the time taken to record the
spectrum without reducing the signal-to-
noise ratio.

T / F

Response

The statements (*i*) and (*iii*) are true.

The statements (*ii*) and (*iv*) are false

If the number of channels is increased then each channel corresponds to a smaller range of energies. It is then possible that two
energy values, E_1 and E_2, previously within the same channel, appear in separate channels. If E_1 and E_2 are in the same channel then
it is not possible to tell them apart (unresolved), but once they are

in different channels the counts are recorded separately making it possible to distinguish between them and *resolve* the two energies. Hence (*i*) is true.

For the S/N ratio of the spectrum we must be concerned with the statistical variation in *each* channel count. By increasing the number of channels the number of counts per channel in the same time expected must decrease. A decrease in count, N_0, produces a *decrease* in the S/N ratio of the count. See Section 4.2.5.

Hence (*iii*) is true and (*ii*) is false.

The S/N ratio is reduced by increasing the number of channels, and reducing the time of the count also reduces the S/N ratio of the count. Hence it is not possible, both to halve the available time, and to increase the number of channels, without expecting a reduction in the S/N ratio.

Hence (*iv*) is false.

In fact, if the number of channels is doubled, then we would have also to double the count time if we were to expect to maintain the same S/N ratio on each channel count.

****

SAQ 4.2b

Detection of a particular beam of X-rays gives a count rate of 40 counts s^{-1}.

Calculate the percentage error (by using one standard deviation) in the result for a counting period of 10 s. $\longrightarrow$

SAQ 4.2b (cont.)	What counting period would be required to reduce the percentage error to one half of this value? (You should be able to do the second part of this question very quickly without very much calculation.)

Response

If the count rate is 40 counts s^{-1}, and the counting period is 10 s, then we would expect that the total count would be 40 x 10 = 400 counts.

If we then put N_0 = 400, we can find the standard deviation from the square root of N_0, ie

$$s = \sqrt{400} = 20.$$

The percentage error is given by 20/400 × 100% = 5% If you have a problem with this refer back to Section 4.2.5.

We know from the text that the S/N ratio in a statistical count improves as the square root of the increase in the counting period. A reduction of the percentage error to half is equivalent to a doubling of the S/N ratio. Thus we must increase the time of the count by a factor equal to 2^2 = 4.

We must increase the counting period by a factor of 4.

If you have a problem with this last part, then repeat the problem in the first part for a period of 4 × 10 = 40 s, and check to see whether it really does give the necessary reduction in the percentage error.

****

SAQ 4.3a   We wish to convert the chromatogram signal, sketched in Fig. 4.3e, into an 8-bit digital signal, the only ATD converter available having an input range of 1.0 V.

(*i*) Calculate the possible percentage error due to conversion in recording each peak.

(*ii*) What would be required of an amplifier if we were to use it to reduce the errors arising because of quantisation?

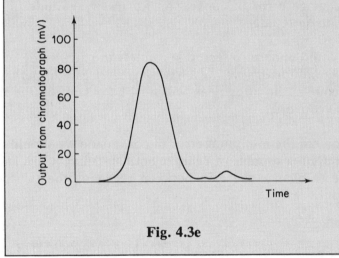

Fig. 4.3e

Response

For a range of 1000 mV, the Quantisation Error of Conversion for an 8-bit ATD converter is equivalent to about 4.0 mV at the input. The magnitudes of the two peaks in the signal are approximately 85 mV and 6 mV.

Thus the percentage error in the major peak = 4.0/85 × 100% ≈ 5%.

The percentage error in the minor peak = 4.0/6.0 × 100% = 67%. This is so large as to make the reading almost valueless.

For efficient conversion into a digital signal with an ATD converter of range 1000 mV we would need to amplify the large signal to about 850 mV. If we amplified the signal to 1000 mV there would be a danger of it going over range if the signal amplitude changed slightly. The appropriate gain for the large signal is therefore about *10*.

We must assume that if the amplifier has a gain of 10 then the small peak will acquire only an amplitude of 6 mV × 10 ≃ 60 mV. The percentage error of conversion for this magnitude of voltage will still be quite large.

We could choose to record separately the amplitude of the smaller peak, and for this we must have a greater gain of perhaps *100–130*. Of course, the large peak will now be cut-off by the range of the ATD converter.

To obtain the minimum error of conversion we should thus require an amplifier capable of a gain of both × 10 and × 100, and we should have to change the gain between receiving the two peaks at the output.

****

SAQ 4.3b	Explain what would be the effect on the electrocardiogram (heart-beat signal) given in Fig. 3.1n if it were fed (with correct amplitude) into a DTA converter with a sampling range of 100 Hz.

Response

The overall period if the electrocardiogram signal as shown is about 0.3 s = 300 ms. The sampling period is about 10 ms (= 1/100 Hz).

Thus for the P and the T waves there are about 6 recording points, and, since they are both smooth waveforms, this will allow a reasonable record of their shape and amplitude to be obtained.

However the R pulse, although of approximately the same width as P and T, has a more definite (triangular) shape. Six recording points for the R pulse will give only an approximation of the true shape, ie some distortion will occur.

The sharp pulses, Q and S, have each a period of only about 10 ms and this allows for perhaps only one or two sampling points, quite insufficient to record even the amplitude of the pulses accurately.

The result is the computer will have a good record of the shape of the P and T pulses, some distortion of the R pulse, and perhaps one or two apparently spurious, readings. The information contained in the sharp peaks would be lost.

You may not have obtained exactly the answer given above, but you should have realised that narrow lines (or pulses), such as Q and S, would be lost, and also that some distortion may occur for other lines.

****

SAQ 4.4a	A chart recorder has possible voltage input ranges of 10.0 mV, 100 mV and 1.00 V, and a chart width of 20.0 cm.
	Express the maximum sensitivity of this recorder in units of cm mV^{-1}.

Response

The most sensitive setting is that with the lowest input-voltage range – 10 mV. With this setting the pen will move right across the paper (20.0 cm) for an input voltage of 10 mV.

The sensitivity, S, is defined as Response/Input.

Hence $S = 20.0/10.0 = 2.0$ cm mV^{-1}

You may have obtained a value of 0.5 by taking the reciprocal of S. This still gives a measure of the sensitivity (albeit as a reciprocal), but then the units would be mV cm^{-1}. This figure would actually decrease for an increase in sensitivity.

If you used one of the other ranges for your calculation then you need to check back into the text to establish the relationship between sensitivity and input range.

****

SAQ 4.4b	A $5\frac{1}{2}$ digit display system has an input voltage range of -2.0 V to $+2.0$ V. What change of input voltage will cause the least significant digit to change by one unit?

Response

The maximum and minimum displays are:

199999 corresponding to $+2.0$ V,

-199999 corresponding to -2.0 V.

The difference between 1.99999 and 2.0 can be ignored because 2.0 is expressed only to an accuracy of one decimal place.

A reading of 100000 must correspond to an input voltage of 1.0 V.

Thus a reading of 000001 must correspond to an input voltage of 0.00001 V $= 0.01$ mV.

A change of 0.01 mV at the input will change the final digit by '1'.

$$************************************$$

SAQ 5.1a A spectrophotometer uses the double-beam *optical-null* principle described in the text. The detector developes a fault which means that its response becomes non-linear. The form of this response is shown in Fig. 5.1e.

Will this fault mean that when the apparent value of absorbance is recorded by the *optical-null* method in the instrument it is:

(*i*)   greater than,
(*ii*) equal to,
(*iii*) less than,

the true value of the absorbance of the sample?

Indicate which option is correct.

Response

The correct answer is (*ii*).

You may find this confusing, particularly as we saw in the text (Section 5.1.4) that the linearity of the detector *does* affect the reading of the *ratio-recording* instrument.

The explanation is based on the fact that the optical-null system is a null-balance measurement method see Section 2.4.3. The detector gives a signal-output to the balance-drive motor (f_1 in the question in Section 5.1.3) *only* when the system is off-balance, and it is this signal which is used to drive the system towards the balance-point. At the actual balance-point there is *no* signal output, f_1, from the detector. Hence the linearity of the detector is irrelevant and will not affect the position of the balance-point.

****

SAQ 5.1b

We wish to re-design a double-beam spectrophotometer so that we can *increase* the frequency of the chopper. As a result of this, which of the following statements will be true?

(*i*) The intensity of the source must be increased.

T / F

(*ii*) The intensity of the source must be decreased.

T / F

(*iii*) The response-time of the detection system must be increased.

T / F

(*iv*) The response-time of the detection system must be decreased.

T / F

Response

The correct answer is:

(*i*)   F

(*ii*)   F

(*iii*)   F

(*iv*)   T.

The frequency of the chopper does not affect the average light intensity reaching the detector. Hence there is no requirement either to increase or to decrease the intensity of the light source.

The concept of response-time was discussed in Section 3.2.5. When the chopper frequency is increased, the *time* between each of the pulses, S and R, will decrease.

If the detector is to respond individually to the pulses it must therefore be able to respond in a shorter time.

Hence its response time must be *decreased*.

If you are confused by this, refer back to Section 2.1.9 and the relationship between frequency and period.

****

SAQ 5.2a   Indicate which of the following statements about the slope-control of the pH meter are true.

If the slope-control is changed from 100% to 80%, the effect is: $\longrightarrow$

SAQ 5.2a
(cont.)

(*i*)   to decrease the input impedance of the instrument.

T / F

(*ii*) to increase the input impedance of the instrument.

T / F

(*iii*) to decrease the gain of the amplifier.

T / F

(*iv*) to increase the gain of the amplifier.

T / F

Response

The answer is:

(*i*)   F

(*ii*)   F

(*iii*) F

(*iv*)   T

The input impedance of a pH-meter should be as large as possible. Provided that the input impedance is far larger than the output impedance of the electrode (see Section 2.3) its exact value will not affect the result. There are no controls on the pH-meter which alter the input impedance of the instrument. Both (*i*) and (*ii*) are false.

We saw in the text that a decrease of slope from 100% to 80% corresponds to a loss in the pH sensitivity of the electrode. If the meter is to give the correct output reading, it is necessary that the gain of the amplifier be *increased* to compensate for the loss in sensitivity of the electrode.

Hence, (*iv*) is the only correct answer.

****

SAQ 5.2b	A computer-memory system as described in Section 5.2.4 is used to reduce the effect of various types of drift in instruments. A list of several forms of drift is given below. Indicate by ticking the 'Y', the types of drift whose effect can be reduced in this way.

Spectrophotometer.

(*i*) Variation of light-source intensity with mains voltage.

Y / N

(*ii*) Variation of detector sensitivity with wavelength.

Y / N

(*iii*) Variation of amplifier-gain with time.

Y / N

Gas Chromatograph.

(*iv*) Variation of detector sensitivity with flow-rate,

Y / N

⟶

SAQ 5.2b
(cont.)
	(*v*) Variation of the physical properties of the column with temperature,
	
	                                                  Y / N
	
	(*vi*) Variation of the flow-rate controller with time.
	
	                                                  Y / N

Response

The correct answer is:

(*i*)   N

(*ii*)   Y

(*iii*)   N

(*iv*)   Y

(*v*)   Y

(*vi*)   N

The important point to remember in this question is that the computer 'baseline' memory-system is only useful for eliminating variations that are *reproducible* from one experimental sweep to the next.

For the spectrophotometer, the only variable that changes reproducibly each time the experiment is repeated is the *wavelength*. The spectrophotometer will normally sweep from one end of its wavelength range to the other. The sensitivity of the detector (*ii*) will change with wavelength, but it will do it in the same way each time the sweep is performed. Hence the computer can be used to memorise any drift due to this variation.

In the gas chromatograph the varying parameter is the *temperature* of the oven as it follows its temperature programme. We would then expect the temperature-dependent properties of the column to be reproducible, and that their effect would be compensated by the use of the memory system. We also expect that in the gas chromatograph, the major cause of a change in flow-rate would be the changing properties of the column. If the properties of the column depend only on temperature, then we should also expect that the flow-rate in the system would also be a function of temperature. The variation of the sensitivity of the detector with flow-rate would then be reproducible from experiment to experiment, and hence could be memorised by the computer.

The other variations mentioned in the question are with respect to changes in time and mains voltage. Any variations of the mains voltage are random, and do not recur reproducibly every time we run an experiment. Similarly if the properties of something are changing with time, either in an ageing process, or in random fluctuations, those changes will not necessarily repeat themselves when we repeat the experiment.

In both these situations, the memory of the computer system cannot help in predicting the values present in a subsequent experiment.

****

SAQ 5.3a

The Specifications for a particular Y-t Chart Recorder contain the following entries.

Chart Width   200 mm

Writing Speed   400 mm . s^{-1}

$\longrightarrow$

SAQ 5.3a
(cont.)

Ranges	$20, 50, 100, 200 \mu V \cdot mm^{-1}$
	$0.5, 1, 2, 5, 10$
	$20, 50, 100, 200$
	$500 \qquad mV \cdot mm^{-1}$
Accuracy at FSD	$\pm 0.2\%$
Linearity	0.1% FSD
Repeatability	0.1% FSD
Input Impedance	1 M ohm on all ranges
Chart Drive Speeds	$2, 5, 10, 20 \ mm \cdot min^{-1}$
	$1, 2, 5, 10, 20, \ mm \cdot s^{-1}$
Zero Setting	Adjustable within full scale.

(i) Estimate the frequency band-width (maximum frequency response) for a varying signal which has an almost full scale deflection.

(ii) Calculate a value for the maximum sensitivity of the instrument.

(iii) What is the possible error (in mV) due to non-linearity on the $1 \ mV \cdot mm^{-1}$ range?

(iv) If the minimum detectable signal is equivalent to 0.2% FSD, calculate the dynamic range of the instrument. $\longrightarrow$

SAQ 5.3a (cont.)	(*v*)	A signal is drawn out with a line-width of 16 mm when the speed of the chart paper is 10 mm . s^{-1}. Calculate the band-width of the signal in the frequency domain.
	(*vi*)	Estimate the possible error, arising from relative input and output impedances, if the recorder is used to measure a voltage originating from a source with an output impedance of 10 kΩ.
	(*vii*)	What is the maximum output impedance of a measured source voltage if the resultant error, as in (*vi*), is to be less than the other total errors in the recorder.

Response

(*i*) Answer: Band-width = 0.8 Hz.

The response-time of the recorder must be calculated from the *writing speed* (400 mm s^{-1}) and the amplitude of the signal (which we are told is almost full scale, ie 200 mm).

Referring to Section 4.4.4, the time taken for the pen to move 90% of the full scale deflection (ie 90% of 200 mm) = 180/400 = 0.45 s.

Thus the response time, as defined in Section 4.4.4, is given below.

$$\tau = 0.45 \text{ s}.$$

The frequency bandwidth is calculated using the relationship given in Section 3.2.5.

$$\Delta f = 1/(2.7 \times \tau) \approx 0.8 \text{ Hz}$$

(*ii*) Answer: Sensitivity = 0.05 mm μV^{-1} = 50 mm mV^{-1}.

The maximum sensitivity will occur with the input set to the smallest range, ie 20 μV mm^{-1}. Hence an input of 20 μV gives a deflection of 1.0 mm.

Sensitivity is defined as Output divided by Input, thus,

$$\text{Sensitivity} = 1.0 \text{ mm}/20 \ \mu V = 0.05 \text{ mm } \mu V^{-1}.$$

If you have problems refer to SAQ 4.4a.

(*c*) Answer: Error = 0.2 mV

On the 1 mV mm^{-1} range, a full-scale deflection is equivalent to 200 mV. The error arising from non-linearity is given below,

$$\text{Error} = \text{Range} \times \text{Linearity} = 200 \times 0.1/100 \text{ mV}$$

$$= 0.2 \text{ mV}.$$

(*iv*) Answer: Dynamic Range = 12.5 × 10^6

The minimum detectable signal will be observed on the most sensitive range as a deflection of 0.2% FSD. We know the FSD is equivalent to 200 × 20 μV = 4000 μV on the most sensitive range, thus the minimum detectable signal, Q(min) = 4000 × 0.2/100 = 8.0 μV.

The maximum measurable signal Q(max), is equivalent to the signal giving a FSD on the least sensitive range, ie 200 × 500 = 100000 mV = 100 V.

The dynamic range is given by the ratio,

$$Q(\text{max})/Q(\text{min}),$$

$$\text{ie } 100/8 \times 10^{-6}$$

$$= 12.5 \times 10^6.$$

(*v*) Answer: Band-width = 0.3 Hz.

At a speed of 10 mm s^{-1} the chart paper will cover a distance of 16 mm in 16/10 = 1.6 s. Thus the width in time of the signal, Δt = 1.6 s.

Thus, from Section 3.1.6, (Eqn. 3.5) the frequency band-width is given by

$$\Delta f = 0.5/1.6 = 0.3 \text{ Hz.}$$

Thus we can see that the band-width of the signal (0.3 Hz) is less than the band-width (0.8 Hz) of the recorder. Hence we do not expect much distortion in the shape of the signal as it was drawn out.

(*vi*) Answer: Error = 1%.

We must assume that the voltage source is acting as a Thévenin circuit (see Section 2.3) with a voltage, V_o, and an output impedance, R_o, = 10 kΩ.

When this is connected to the recorder, the combined resistance of the circuit,

$$R_o + R_i = 10 \text{ } k\Omega + 1M\Omega = 1010 \text{ } k\Omega,$$

where R_i is the input impedance of the detector, see Fig. 2.3i.

Thus the current flowing into the recorder, I, is given by.

$$I = V_o/R = V_o/1010 \times 10^3 \text{ A.}$$

The resultant potential differences across the input to the recorder;

$$V = I. \times R_i$$

$$= V_o \times 1000 \times 10^3/1010 \times 10^3$$

$$= \frac{1}{1.01} \times V_o = 0.99 \text{ } V_o.$$

Thus the measured voltage, V, is 1% less than the true voltage, V_0.

If you find this difficult to understand, refer back to SAQ 2.3a.

(*vii*) Answer: Maximum output impedance = 2 kΩ.

The general equation, that relates the measured voltage, V, to true voltage, V_0, is as follows.

$$V = V_0 \times R_i/(R_0 + R_i).$$

This was, in fact, the calculation used in (*vi*).

We saw in SAQ 2.3a that provided $R_i \gg R_0$, the percentage error in measurement is given by,

$$\text{Error } (\%) = R_0/R_i \times 100.$$

In this question, the total error (=accuracy) of the recorder is 0.2%. Thus

$$R_0 = \frac{0.2}{100} \times 1 \times 10^6 \text{ ohm} = 2 \text{ kΩ}$$

****

SAQ 5.3b

The Specifications for an Infra-red Spectrophotometer contain the following entries.

Principle	Double-beam optical-null
Digital Display	4 digit LED display of wavenumber
Abscissa Ranges	4000–2000 cm^{-1},
	2000–200 cm^{-1},
	running consecutively.

$\longrightarrow$

SAQ 5.3b (cont.)

| Abscissa Accuracy | $4000–2000, \pm 6 \text{ cm}^{-1}$ |
| | $2000–200, \pm 3 \text{ cm}^{-1}$ |

Abscissa Repeatability Within 0.6 mm of chart

S/N Better than $100:1$, outside the region of atmospheric absorption and with narrow slit programme.

Resolution Narrow Slit Programme 3 cm^{-1} at 1000 cm^{-1} 5 cm^{-1} at 3000 cm^{-1}

Chart Size $15 \text{ cm} \times 47 \text{ cm}$ (see Fig. 5.3n.)

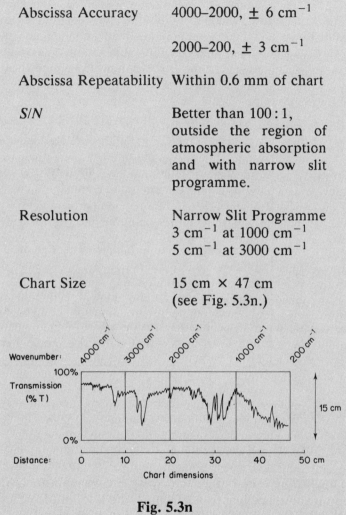

Fig. 5.3n

Note. In Infra-red spectroscopy, the 'wavelength' scale is not calibrated in nanometres. The scale is actually *inversely* proportional to the wavelength, and is calibrated in 'wavenumbers' which have units of cm^{-1} and are obtained by taking the reciprocal of the wavelength expressed in centimetres; $\longrightarrow$

SAQ 5.3b
(cont.)

Wavenumber (cm^{-1}) = 1/Wavelength (cm)

It is also common, for technical reasons, for the wavenumber scale in an ir spectrophotometer to be divided into two ranges.

(*i*)   What is the wavenumber readability in cm^{-1} on the digital display?

(*ii*)   Calculate the repeatability in cm^{-1} for the two ranges on the chart.

(*iii*) In which range does the instrument have the better resolution?

(*iv*)   Why does the signal-to-noise ratio decrease if there is absorption of radiation in the atmosphere (which would affect both beams similarly), or if a narrow-slit programme is used in the monochromator?

Response

(*i*) Answer: Readability = $\pm$ 1 cm^{-1}.

The display uses 4 digits to display the wavenumber. Since the maximum value is equal to 4000 cm^{-1}, the last, and least significant digit, therefore corresponds to 1 cm^{-1}. See Section 5.3.5.

(*ii*) Answer: Repeatability = 6 cm^{-1}, 4 cm^{-1}.

The repeatability is given as a range of 0.6 mm on the chart. We can see from Fig. 5.3n, that the two wavenumber ranges have different scale factors.

The scale 4000 – 2000 cm^{-1} covers a distance on the chart of 20 cm.

Thus

$$20 \text{ cm (chart)} \equiv 4000 - 2000 = 2000 \text{ cm}^{-1}$$

$$0.6 \text{ mm (chart)} \equiv \frac{0.6}{200} \times 2000^{-1}.$$

$$= 6 \text{ cm}^{-1}$$

The scale 2000 to 200 cm^{-1} covers a distance of 27 cm.

Thus,

$$27 \text{ cm (chart)} = 2000 - 200 = 1800 \text{ cm}^{-1}$$

$$0.6 \text{ mm (chart)} = \frac{0.6}{270} \times 1800^{-1}$$

$$= 4 \text{ cm}^{-1}.$$

(*iii*) Answer: The better resolution is in the range 2000 – 200 cm^{-1}.

The resolution is given as the separation (in cm^{-1}) between two lines which can just be resolved, see Section 5.3.3. Thus, the resolution improves if the minimum separation decreases.

A resolution of 3 cm^{-1} is therefore better than a resolution of 5 cm^{-1}.

The resolution of 3 cm^{-1} occurs at 1000 cm^{-1} which is in the lower wavelength range.

(*iv*) The double-beam spectrophotometer involves a comparative method. A reduction in radiation intensity, which affects both beams similarly does not change the 'balance point' in the system, and, therefore, does not change the measured value of sample absorbance, see Section 5.1.2. This is the situation for a decrease in monochromator slit width or absorption in the air.

However, although the overall signal (optical or electronic) may be reduced, the noise signals (optical and electronic) are not reduced. When the gain of the system is increased to compensate for the decrease in signal intensity (see Section 5.1.3) the noise signals are made larger, and more noise appears on the output of the instrument.

Hence, although there are no new systematic errors introduced, the decrease in signal intensity results in a relative increase of random noise.

****

SAQ 5.3c

The Specifications of a pH meter contain the following entries.

Read-out	$3\frac{1}{2}$ digit display
Ranges	0 to 14.00 pH
	0 to 1900 mV
	−30 to 105 °C
Automatic Temperature Compensation	0 to 100 °C
Manual Temperature Compensation	0 to 100 °C
Slope Correction	80 to 110 %
Accuracy	± 0.06 pH
	± 6 mV

$\longrightarrow$

SAQ 5.3c
(cont.)

Repeatability	± 0.02 pH
	± 2 mV
Linearity	± 0.04 pH
	± 4 mV
Input impedance	$> 10^{13}$ ohms
Recorder Output	$10 \text{ mV} \cdot \text{pH}^{-1}$
Response Time	10 ms

Note that the pH-meter can be used either to record pH directly, or to record the voltage (in mVs) from other types of electrodes. It can also be used to measure temperature.

(i) If the input impedance could be increased by a factor of 10 from 10^{13} to 10^{14} ohms, by what factor would the accuracy of the instrument change?

(ii) If the instrument is calibrated at pH 4 and pH 9, estimate the error due to non-linearity in a reading at a value of about pH 6.5.

(iii) Is it essential to use an automatic temperature probe with this instrument?

(iv) If we wish to display the full pH range (0 to 14) on the recorder described in SAQ 5.3a, which range should be selected for the recorder?   ⟶

SAQ 5.3c
(cont.)

(v)   If we were to connect the pH-meter to the recorder detailed in SAQ 5.3a, is the performance of the recorder plus pH-meter system significantly worse than the performance of the pH-meter on its own?

If so, in which respect is it worse?

(iv) Is it possible with this instrument to correct the pH-meter if its sensitivity to pH has decreased due to ageing?

Response

(i) The accuracy of the instrument will not change if the input impedance of the pH meter is increased. We saw from SAQ 2.3a that the error in measurement due to a non-infinite input impedance is

$$\text{Error} = R_o/R_i \times 100\%$$

where R_o is the output impedance of the pH-electrode, and R_i is the input impedance of the pH-meter. If we take the output impedance of the pH-electrode to be as high as 100 MΩ (see Section 2.3.2), then the error is as follows.

$$\text{Error} = \frac{(100 \times 10^6)}{10^{13}} \, 1 \times 100$$

$$= 0.001\%.$$

which is equivalent to a maximum error of

$$\frac{0.001}{100} \times 14 = \pm 0.00014 \text{ pH units.}$$

This error is insignificant compared with the total errors in the instrument,

$$= \pm 0.06 \text{ pH units.}$$

Thus increasing the input impedance can have no significant effect in reducing this total error which is due to other causes.

In addition to the reasons given above, we also know that slight errors due to a systematic error in the measurement of the voltage can be compensated for in the operation of the pH-meter. The act of calibration against *two* standards using the *slope*-control will automatically allow for a lack of sensitivity of the pH electrode. It will also allow for a constant proportional error in the measurement of its emf.

(*ii*) The error due to linearity is given as 0.04 pH over a range of 14 pH units. If we standardize with pH values of 4.00 pH and 9.00 pH, we reduce the range of measurement to $9.00 - 4.00 = 5.00$ pH units. This is a reduction in range by a factor of $14/5 = 2.8$. The error due to non-linearity will be reduced by a factor equal to the square of this ratio $= (2.8)^2 = 7.84$.

Thus the error due to non-linearity at pH 6.5, which is mid-way between pH 4.00 and pH 9.00, may be

$$\pm \ \frac{0.04}{7.84} \simeq \pm \ 0.005 \text{ pH unit.}$$

We have seen that by making calibration points close to measured values, the errors due to non-linearity are greatly reduced.

(*iii*) The answer is *no*.

Although there is a facility for an automatic temperature compensation, there is also the capability for *manual* temperature compensation. Provided that the temperature of the solution can be found by some other means (eg a thermometer), then a special probe is not required.

(*iv*) The recorder output from the pH-meter is specified as 10 mV pH^{-1}. Thus over the range of the meter, pH 0 to pH 14, the voltage output covers the range from 0 V to $14 \times 10 = 140$ mV.

The recorder has ranges of 0.5 mV mm^{-1} and 1 mV mm^{-1}, together with a chart width of 200 mm. These two ranges give a FSD for inputs of (0.5 × 200 mV =) 100 mV and (1 × 200 mV =) 200 mV respectively. The 0.5 mV mm^{-1} setting will not allow the complete pH range (0 to 14) to be displayed.

Thus the range with the greatest sensitivity which can still display the complete pH range, is 1 mV mm^{-1}.

(*v*) We can compare the performance of the two instruments in terms of the following percentages of FSD.

	Recorder	pH-meter
Accuracy	0.2%	0.06/14.00 × 100 ≃ 0.4 %
Repeatability	0.1%	0.02/14.00 × 100 ≃ 0.15%
Linearity	0.1%	0.04/14.00 × 100 ≃ 0.3 %

The errors in the recorder are less than those in the pH meter. When they are combined together in a complete system the total error will not be very much larger than the total error already existing in the pH-meter.

What about the effects of output and input impedance? We calculated in SAQ 5.3a(*vii*) that, to keep errors below 0.2%, the output impedance of the source voltage (in this case the pH-meter) must be less than 2 KΩ. We are not given the output impedance of the 'recorder output' from the pH meter, but it is probable that it is only of the order of 1 kΩ. However, any possible error here can be eliminated by calibrating the pH-meter by using the *slope* control so that the recorder reads the *correct* values at two calibrating points – see the answer to (*i*).

There is also one other factor which we have not yet discussed: the response time. We calculated in SAQ 5.3a(*i*) that the response time of the recorder is 0.45 s = 450 ms. This is considerably *longer* than the response-time of the pH-meter, 10 ms. Hence, connecting the recorder to the pH-meter has increased the response time for a measurement from 10 ms to about 450 ms. For most measurements, however, this is irrelevant because the response time of most pH-electrodes is even longer than this. Nevertheless, there may be occasions in measurement of a mV signal when the pH meter is expected to respond more quickly. The limiting effect of the response-time of the recorder should be remembered.

(*vi*) The answer is *yes*.

This pH meter has a *slope*-control which can be used to increase the gain of the amplifier to compensate for a decrease in sensitivity of the pH-electrode, see Section 5.2.2.

Units of Measurement

For historic reasons a number of different units of measurement have evolved to express quantity of the same thing. In the 1960s, many international scientific bodies recommended the standardisation of names and symbols and the adoption universally of a coherent set of units—the SI units (Système Internationale d'Unités)—based on the definition of five basic units: metre (m); kilogram (kg); second (s); ampere (A); mole (mol); and candela (cd).

The earlier literature references and some of the older text books, naturally use the older units. Even now many practicing scientists have not adopted the SI unit as their working unit. It is therefore necessary to know of the older units and be able to interconvert with SI units.

In this series of texts SI units are used as standard practice. However in areas of activity where their use has not become general practice, eg biologically based laboratories, the earlier defined units are used. This is explained in the study guide to each unit.

Table 1 shows some symbols and abbreviations commonly used in analytical chemistry. Table 2 shows some of the alternative methods for expressing the values of physical quantities and the relationship to the value in SI units.

More details and definition of other units may be found in the *Manual of Symbols and Terminology for Physicochemical Quantities and Units*, Whiffen, 1979, Pergamon Press.

Table 1 *Symbols and Abbreviations Commonly used in Analytical Chemistry*

Å	Angstrom
$A_r(X)$	relative atomic mass of X
A	ampere
E or U	energy
G	Gibbs free energy (function)
H	enthalpy
J	joule
K	kelvin $(273.15 + t\,°C)$
K	equilibrium constant (with subscripts p, c, therm etc.)
K_a, K_b	acid and base ionisation constants
$M_r(X)$	relative molecular mass of X
N	newton (SI unit of force)
P	total pressure
s	standard deviation
T	temperature/K
V	volume
V	volt $(J\ A^{-1}\ s^{-1})$
$a, a(A)$	activity, activity of A
c	concentration/ mol dm^{-3}
e	electron
g	gramme
i	current
s	second
t	temperature / °C
bp	boiling point
fp	freezing point
mp	melting point
$\approx$	approximately equal to
$<$	·less than
$>$	greater than
e, $\exp(x)$	exponential of x
$\ln x$	natural logarithm of x; $\ln x = 2.303 \log x$
$\log x$	common logarithm of x to base 10

Table 2 *Alternative Methods of Expressing Various Physical Quantities*

1. **Mass (SI unit : kg)**

$$g = 10^{-3} \text{ kg}$$
$$mg = 10^{-3} \text{ g} = 10^{-6} \text{ kg}$$
$$\mu g = 10^{-6} \text{ g} = 10^{-9} \text{ kg}$$

2. **Length (SI unit : m)**

$$cm = 10^{-2} \text{ m}$$
$$\text{Å} = 10^{-10} \text{ m}$$
$$nm = 10^{-9} \text{ m} = 10\text{Å}$$
$$pm = 10^{-12} \text{ m} = 10^{-2} \text{ Å}$$

3. **Volume (SI unit : m^3)**

$$l = dm^3 = 10^{-3} \text{ m}^3$$
$$ml = cm^3 = 10^{-6} \text{ m}^3$$
$$\mu l = 10^{-3} \text{ cm}^3$$

4. **Concentration (SI units : mol m^{-3})**

$$M = \text{mol } l^{-1} = \text{mol dm}^{-3} = 10^3 \text{ mol m}^{-3}$$
$$\text{mg } l^{-1} = \mu g \text{ cm}^{-3} = ppm = 10^{-3} \text{ g dm}^{-3}$$
$$\mu g \text{ g}^{-1} = ppm = 10^{-6} \text{ g g}^{-1}$$
$$ng \text{ cm}^{-3} = 10^{-6} \text{ g dm}^{-3}$$
$$ng \text{ dm}^{-3} = pg \text{ cm}^{-3}$$
$$pg \text{ g}^{-1} = ppb = 10^{-12} \text{ g g}^{-1}$$
$$mg\% = 10^{-2} \text{ g dm}^{-3}$$
$$\mu g\% = 10^{-5} \text{ g dm}^{-3}$$

5. **Pressure (SI unit : N m^{-2} = kg m^{-1} s^{-2})**

$$Pa = Nm^{-2}$$
$$atmos = 101\ 325 \text{ N m}^{-2}$$
$$bar = 10^5 \text{ N m}^{-2}$$
$$torr = mmHg = 133.322 \text{ N m}^{-2}$$

6. **Energy (SI unit : J = kg m^2 s^{-2})**

$$cal = 4.184 \text{ J}$$
$$erg = 10^{-7} \text{ J}$$
$$eV = 1.602 \times 10^{-19} \text{ J}$$

Table 3   *Prefixes for SI Units*

Fraction	Prefix	Symbol
10^{-1}	deci	d
10^{-2}	centi	c
10^{-3}	milli	m
10^{-6}	micro	μ
10^{-9}	nano	n
10^{-12}	pico	p
10^{-15}	femto	f
10^{-18}	atto	a

Multiple	Prefix	Symbol
10	deka	da
10^2	hecto	h
10^3	kilo	k
10^6	mega	M
10^9	giga	G
10^{12}	tera	T
10^{15}	peta	P
10^{18}	exa	E

Table 4   *Recommended Values of Physical Constants*

Physical constant	Symbol	Value
acceleration due to gravity	g	9.81 m s^{-2}
Avogadro constant	N_A	$6.022\ 05 \times 10^{23} \text{ mol}^{-1}$
Boltzmann constant	k	$1.380\ 66 \times 10^{-23} \text{ J K}^{-1}$
charge to mass ratio	e/m	$1.758\ 796 \times 10^{11} \text{ C kg}^{-1}$
electronic charge	e	$1.602\ 19 \times 10^{-19} \text{ C}$
Faraday constant	F	$9.648\ 46 \times 10^{4} \text{ C mol}^{-1}$
gas constant	R	$8.314 \text{ J K}^{-1} \text{ mol}^{-1}$
'ice-point' temperature	T_{ice}	$273.150 \text{ K exactly}$
molar volume of ideal gas (stp)	V_m	$2.241\ 38 \times 10^{-2} \text{ m}^3 \text{ mol}^{-1}$
permittivity of a vacuum	ϵ_0	$8.854\ 188 \times 10^{-12} \text{ kg}^{-1}$ $\text{m}^{-3} \text{ s}^4 \text{ A}^2 \text{ (F m}^{-1})$
Planck constant	h	$6.626\ 2 \times 10^{-34} \text{ J s}$
standard atmosphere pressure	p	$101\ 325 \text{ N m}^{-2} \text{ exactly}$
atomic mass unit	m_u	$1.660\ 566 \times 10^{-27} \text{ kg}$
speed of light in a vacuum	c	$2.997\ 925 \times 10^{8} \text{ m s}^{-1}$